Lecture Notes in Statistics

Volume 228

Series Editors

Peter Diggle, Department of Mathematics, Lancaster University, Lancaster, UK

Scott Zeger, Baltimore, USA

Ursula Gather, Dortmund, Germany

Peter Bühlmann, Seminar for Statistics, ETH Zürich, Zürich, Switzerland

Lecture Notes in Statistics (LNS) includes research work on topics that are more specialized than volumes in Springer Series in Statistics. The series seeks to publish a broad portfolio of scientific books, aiming at researchers and students. The series includes peer-reviewed monographs, textbooks and lecture notes.

The series editors are currently Peter Bühlmann, Peter Diggle, Ursula Gather.

Alain Franc

Linear Dimensionality Reduction

 Springer

Alain Franc
INRAE BioGeCo & INRIA Pleiade
Cestas, France

ISSN 0930-0325   ISSN 2197-7186   (electronic)
Lecture Notes in Statistics
ISBN 978-3-031-95784-0   ISBN 978-3-031-95785-7   (eBook)
https://doi.org/10.1007/978-3-031-95785-7

Preface

The writing of this document was motivated by the convergence of several observations.

◇ The diversity of methods described as Data Analysis in the 1970s, nowadays attached to Statistical Learning, or Data Mining, can be organized in a circular way where one method is a variation of another via certain choices. This global approach is essentially algebraic, based on matrix calculation. It has been developed in particular by a French school in parallel with more statistical approaches developed in Anglo-Saxon countries.

◇ The main methods are PCA (Principal Component Analysis) associating items and features, CoA (Correspondence Analysis) of contingency tables, Canonical Analysis for analysis of two tables, Multiple Correspondence Analysis for analysis of two or more tables, MDS (Multidimensional Scaling) for building a point cloud from a matrix of distances, etc. These methods had fallen into disuse as descriptive methods, but have experienced a revival of interest in exploring structures in massive data (which is sometimes called "pattern discovery" and is related to unsupervised learning), mainly because statistical learning methods are sometimes inoperative in spaces of very large dimension. A pretreatment as a projection on a space of lower dimension can make them operational, provided the projection respects the structure of the dataset.

◇ The starting point for each of these methods is a data array, which can be a cross-tabulation between objects and variables, a contingency table, a distance or dissimilarity matrix, etc. A key observation is that each of the methods is organized according to the triptych:

where each method can be read as a diabolo, where pre-processing of the data constructs a matrix, which is itself decomposed via a Singular Value Decomposition (SVD), the outputs of which are post-processed to produce the desired result. The SVD step is generally the limiting step for scaling up, i.e., processing massive data given as very large arrays: the complexity is cubic with the size of the arrays.

Significant progress has been made recently in scaling up the SVD by combining three elements:

- An evolution of the SVD computation algorithm by including the Gaussian Random Projection (rSVD, see [BM01, HMT11])
- A distributed memory implementation of the basic matrix operations
- The use of a task-based programming paradigm for the assembly of these steps

Random projection itself is often presented as a dimensionality reduction method, by projecting a point cloud on a space of lower dimension while respecting the pairwise distances (see e.g., [BM01]). In this book, it is used as a tool for lowering the complexity of the calculation of the SVD of a large matrix, as in [HMT11]. The integration of the rSVD in the MDS algorithm has been made in [Bla17, Par18, BCF$^+$18]. Numerical implementation for very large matrices (namely $10^6 \times 10^6$) with distributed-memory and task-based programming has been made in [ACD$^+$22]. A motivation for this book was to show how such an approach developed for MDS can be operational for any linear dimensionality reduction method that relies on an SVD.

About Notes and References When possible we have included some notes and bibliographic references on several topics, sometimes with an historical taste. Indeed, Linear Dimensionality Reduction is an established domain whose main ideas and developments are several decades old. These notes serve as an indicator of the maturity of the subject, and by no means are intended to present a history of the domain. They are entry points for the curious reader, nothing more.

Acknowledgments These notes have been written as a methodological companion to two ADTs (Action de Développement Technologique) built on collaborations between INRIA and Inrae at Bordeaux: Gordon (2019–2020) and Diodon (2021–2022), the aim of which was to provide libraries for running linear dimension reduction on massive data sets [ACD+22] with rSVD, possibly (in C++) distributed memory and task-based programming. I am particularly grateful to Emmanuel Agullo, Pierre Blanchard, Olivier Coulaud, Jean-Marc Frigerio, Emmanuel Jeannot, Romain Péressoni, and Florent Pruvost for many discussions throughout these projects and before, especially on linear algebra, and their encouragements to make explicit this methodological companion which enabled the continuation of the Gordon ADT through the Diodon ADT. I am particularly indebted to Francis Cailliez, Daniel Chessel, Jean-Baptiste Denis, Yves Escoufier, Jean-Dominique Lebreton, and Robert Sabatier, who taught me multivariate data analysis many decades ago.

Competing Interests The author has no competing interests to declare that are relevant to the content of this manuscript.

Cestas, France   Alain Franc

Contents

Abbreviations

CCA	Canonical Correlation Analysis
CoA	Correspondence Analysis
FA	Factor Analysis
GRP	Gaussian Random Projection
IV	Instrumental Variables
MCoA	Multiple Correspondence Analysis
MDA	Multivariate Data Analysis
MDS	Multidimensional Scaling
PCA	Principal Component Analysis
PPCA	Probabilistic PCA
RP	Random Projection
rSVD	Randomized SVD
SDP	Symmetric Definite Positive
SGF	Symmetric Gauge Function
SVD	Singular Value Decomposition
UIN	Unitarily Invariant Norm

Notations

The notations adopted in this book are classical notations for linear algebra. Some are used all over the book, and some others are specific to a given section. We have tried to be as consistent as possible between sections, but the diversity of situations and matrices involved, as well as complying with historically well-accepted notations, makes such a challenge sometimes ... challenging.

$\mathbb{N}$ is the set of non-negative integers, and $\mathbb{R}$ the set of reals. We will seldom work in the set $\mathbb{C}$ of complex numbers.

Integers, reals, and vectors are denoted with lowercase latin letters like $a, b, \ldots, i, \ldots$. Matrices are denoted with uppercase roman letters, like $A, B, \ldots$.

The symbol $[\![a, b]\!]$ where $a, b \in \mathbb{N}$ denotes the set of integers $i \in \mathbb{N}$ with $a \leq i \leq b$.

Most of analysis will be developed on a real matrix A with n rows and p columns, which formalizes a data set. The set of these matrices is denoted $\mathbb{R}^{n \times p}$. The rows will be labeled by i with $1 \leq i \leq n$, and the columns by j with $1 \leq j \leq p$. The term in A at row i and column j is denoted a_{ij}. The row i is denoted a_i, and is a vector in $\mathbb{R}^p$. The column j is denoted a_{*j} and is a vector in $\mathbb{R}^n$. The fact that, for a matrix A, all terms are non-negative ($a_{ij} \geq 0$) is denoted $A \geq 0$. The identity matrix in a space of dimension n is denoted $\mathbb{I}_n$.

A matrix A is the expression of a linear map from a vector space $F = \mathbb{R}^p$ on a vector space $E = \mathbb{R}^n$ once a basis has been chosen in each vector space. In abstract linear algebra, a linear map is an element of $\mathcal{L}(F, E)$. In numerical linear algebra, a matrix is an array in $\mathbb{R}^{n \times p}$.

Affine space $\mathbb{R}^p$ containing the origin will be denoted $\mathbb{R}^p$ or E, like a vector space. A point cloud $\mathcal{A}$ in $\mathbb{R}^p$ is attached to A, with n points. There is a one-to-one correspondence between the points in $\mathcal{A}$ and the rows of A, with point i being a_i.

The rank of a matrix is denoted by r when possible. A matrix of rank r is denoted A_r. A vector subspace of $E = \mathbb{R}^p$ of dimension r is denoted E_r. A point cloud in E_r is denoted $\mathcal{A}_r$. The projection of a point cloud on E_r is denoted $\mathcal{A}_E$.

Here is a summary of these notations adopted throughout the book:

$\|.\|$	Frobenius norm (unless otherwise stated)
$\|.\|_{sp}$	spectral norm
$\otimes$	tensor product between vectors
$[\![a, b]\!]$	the set of integers i with $a \leq i \leq b$
a_{ij}	coefficient in row i and column j of matrix A
a_i	point i in point cloud $\mathcal{A}$
	row i of matrix A $(\in \mathbb{R}^p)$
a_{*j}	column j of matrix A $(\in \mathbb{R}^n)$
A	a matrix in $\mathbb{R}^{n \times p}$
A_r	a matrix of rank r
$A \geq 0$	a non-negative matrix
$\mathcal{A}$	a point cloud of n points in $\mathbb{R}^p$
$\mathcal{A}_E$	projection of point cloud $\mathcal{A}$ on subspace E
E_r	a subspace of $\mathbb{R}^p$ of dimension r
$\mathbb{I}_n$	the identity matrix in $\mathbb{R}^{n \times n}$
$\mathcal{L}(E, F)$	the space of linear functions from E to F
n	number of rows of matrix A
p	number of columns of matrix A
r	prescribed rank for best low rank approximation
$\mathbb{R}^{n \times p}$	the space of matrices with n rows and p columns

Notations of Chapter 2

Symbol	in space	Meaning
C	$\mathbb{R}^{p \times p}$	matrix of correlations
λ	$\mathbb{R}$	eigenvalue of C
Λ	$\mathbb{R}^{p \times p}$	diagonal matrix of eigenvalues of C
r	$\mathbb{N}$	rank
σ	$\mathbb{R}$	singular value of A
Σ	$\mathbb{R}^{p \times p}$	diagonal matrix of singular values of A

(continued)

Symbol	in space	Meaning
U	$\mathbb{R}^{n \times p}$	left singular vectors of U, column-wise
v	$\mathbb{R}^{p}$	eigenvector of C (principal axis)
V	$\mathbb{R}^{p \times p}$	matrix of eigenvectors of C, column-wise
		right singular vectors of A, column-wise
Y	$\mathbb{R}^{n \times p}$	matrix of principal components

Notations of Chapter 4

Symbol	in space	Meaning
$\langle ., . \rangle_{\mathrm{N}}$		inner product induced by N in $\mathbb{R}^n$
$\|.\|_{\mathrm{N}}$		norm induced by N in $\mathbb{R}^n$
$\langle ., . \rangle_{\mathrm{P}}$		inner product induced by P in $\mathbb{R}^p$
$\|.\|_{\mathrm{P}}$		norm induced by P in $\mathbb{R}^p$
$\|.\|_{\mathrm{T}}$		norm induced by T in $\mathbb{R}^{n \times p}$
A	$\mathbb{R}^{n \times p}$	matrix to be analyzed
B	$\mathbb{R}^{n \times p}$	a matrix
F	$\subset \mathbb{R}^p$	linear subspace of $\mathbb{R}^p$
Λ	$\mathbb{R}^{p \times p}$	eigenvalues of $R^{\mathrm{T}} R$
M	$\mathbb{R}^{n \times n}$	unique SDP matrix with $M^2 = N$
N	$\mathbb{R}^{n \times n}$	SDP matrix defining an inner product in $\mathbb{R}^n$
$\mathcal{P}_v$	$\mathcal{L}(\mathbb{R}^p)$	projector on $\mathbb{R}v$ in $\mathbb{R}^p$ for inner product P
$\mathcal{P}_{\mathrm{F}}$	$\mathcal{L}(\mathbb{R}^p)$	projector on $F \subset \mathbb{R}^p$ in $\mathbb{R}^p$ for inner product P
P	$\mathbb{R}^{p \times p}$	SDP matrix defining an inner product in $\mathbb{R}^p$
Q	$\mathbb{R}^{p \times p}$	unique SDP matrix with $P = Q^2$
R	$\mathbb{R}^{n \times p}$	matrix for calculations: $R = MAQ$
Σ		see (U', Σ, W)
T	$\mathbb{R}^{np \times np}$	SDP matrix defining an inner product in $\mathbb{R}^{n \times p}$
(U', Σ, W)		SVD of R, $R = U' \Sigma W^{\mathrm{T}}$
V	$\mathbb{R}^{p \times p}$	principal axis of A for inner product defined by (N, P)
w	$\mathbb{R}^p$	weights for an inner product in R^p
X		principal components of PCA of R with standard inner product
Y	$\mathbb{R}^{n \times p}$	principal components of A with inner product defined by (N, P)
Z	$\mathbb{R}^{np \times np}$	unique SDP matrix with $Z^2 = T$; $Z(A) = MAQ$

Notations of Chapter 5

Symbol	in space	Meaning
A	$\mathbb{R}^{I \times J}$	matrix of frequencies (T/n_{++})
a_{ij}	$[0, 1] \subset \mathbb{R}$	general term of A
c	$\mathbb{R}^J$	vector of marginals of A by columns: $(c_1, \ldots, c_J)$
c_j	$\mathbb{R}$	marginal of column j of A: $(\sum_i a_{ij})$
D_c	$\mathbb{R}^{J \times J}$	diagonal matrix with elements c_j
D_r	$\mathbb{R}^{I \times I}$	diagonal matrix with elements r_i
i	$\mathbb{N}$	indices of rows
I	$\mathbb{N}$	number of rows of T
j	$\mathbb{N}$	indices of columns
J	$\mathbb{N}$	number of columns of T
n_{ij}	$\mathbb{N}$	number of item in class i (row) and j (column)
n_{i+}	$\mathbb{N}$	sum of terms in row i of T
n_{+j}	$\mathbb{N}$	sum of terms in column j of T
n_{++}	$\mathbb{N}$	sum of terms in T
r	$\mathbb{R}^I$	vector of marginals of A by rows: $(r_1, \ldots, r_I)$
r_i	$\mathbb{R}$	marginal of row i of A: $(\sum_j a_{ij})$
T	$\mathbb{R}^{I \times J}$	contingency table

Notations of Chapter 6

Symbol	in space	What it is
$\otimes$		tensor product: $y \otimes v = yv^{\mathrm{T}}$
A	$\mathbb{R}^{n \times p}$	matrix to be analyzed
A_r	$\mathbb{R}^{n \times p}$	best approximation of A of rank r with constraints
F	$\subset \mathbb{R}^n$	subspace of constraints on principal components
H	$\subset \mathbb{R}^p$	subspace of constraints on principal axis
Λ	$\mathbb{R}^{q \times q}$	diagonal matrix of eigenvalues of PCA-IV
m	$\mathbb{N}$	dimension of F
P_{F}	$\mathbb{R}^{n \times n}$	projector on F in $\mathbb{R}^n$
P_{H}	$\mathbb{R}^{p \times p}$	projector on H in $\mathbb{R}^p$
$\mathcal{P}_{\mathrm{F \otimes H}}$	$\mathbb{R}^{np \times np}$	projector on $F \otimes H$
q	$\mathbb{N}$	dimension of H

(continued)

Symbol	in space	What it is
r	$\mathbb{N}$	prescribed rank for best approximation
U_F	$\mathbb{R}^{n \times m}$	matrix of an orthonormal basis of F
V	$\mathbb{R}^{p \times q}$	matrix of principal axis of PCA-IV of (A, F, H)
V_H	$\mathbb{R}^{p \times q}$	matrix of an orthonormal basis of H
V_r	$\mathbb{R}^{p \times r}$	the r first principal axis
Y	$\mathbb{R}^{n \times m}$	principal components of PCA-IV of (A, F, H)
Y_r	$\mathbb{R}^{n \times r}$	first r principal components of PCA-IV of (A, F, H)

Notations of Chapter 7

Symbol	space	Meaning
A	$\mathbb{R}^{n \times p}$	one of the two matrices to be analyzed
B	$\mathbb{R}^{n \times q}$	one of the two matrices to be analyzed
M	$\mathbb{R}^{p \times p}$	$M = N^{1/2}$
N	$\mathbb{R}^{p \times p}$	SDP matrix defining an inner product in $\mathbb{R}^p$; $N = (A^\mathrm{T} A)^{-1}$
P	$\mathbb{R}^{q \times q}$	SDP matrix defining an inner product in $\mathbb{R}^q$; $P = (B^\mathrm{T} B)^{-1}$
Q	$\mathbb{R}^{q \times q}$	$Q = P^{1/2}$
R	$\mathbb{R}^{p \times q}$	for calculation; $R = MTQ$
T	$\mathbb{R}^{p \times q}$	for calculation; $T = A^\mathrm{T} B$
v_A	$\mathbb{R}^p$	a vector in $\mathbb{R}^p$
v_B	$\mathbb{R}^q$	a vector in $\mathbb{R}^q$
V_A	$\mathbb{R}^{p \times q}$	matrix with axis v_A column-wise
V_B	$\mathbb{R}^{q \times q}$	matrix with axis v_B column-wise
y_A	$\mathbb{R}^n$	a vector in span A
y_B	$\mathbb{R}^n$	a vector in span B
Y_A	$\mathbb{R}^{n \times q}$	matrix with components y_A column-wise
Y_B	$\mathbb{R}^{n \times q}$	matrix with components y_B column-wise

Notations of Chapter 8

Symbol	space	Meaning
A	$\mathbb{R}^{n\times p}$	a matrix to be analyzed
A_ℓ	$\mathbb{R}^{n\times p_\ell}$	a matrix to be part of a MCCA
B	$\mathbb{R}^{n\times q}$	a matrix to be analyzed
$(A\quad B)$	$\mathbb{R}^{n\times(p+q)}$	column-wise concatenation of A and B, blockwise notation
C	$\mathbb{R}^{n\times(p+q)}$	$C = (A\quad B)$
	$\mathbb{R}^{n\times s}$	$C = (A_1\ \ldots\ A_m)$, column-wise concatenation of $A_1, \ldots, A_m$
D	$\mathbb{R}^{(p+q)\times(p+q)}$	metrics on $\mathbb{R}^{p+q}$ with N and P as diagonal blocks
	$\mathbb{R}^{s\times s}$	metrics on $\mathbb{R}^s$ with $D_1, \ldots, D_\ell, \ldots, D_m$ as diagonal blocks
λ	$\mathbb{R}$	eigenvalue of PCA of S
M	$\mathbb{R}^{p\times p}$	$M = N^{1/2}$
M_ℓ	$\mathbb{R}^{p_\ell\times p_\ell}$	$M_\ell = N_\ell^{1/2}$
n	$\mathbb{N}$	number of rows in matrices involved in a MCCA $(A, B, A_\ell, \ldots)$
N	$\mathbb{R}^{p\times p}$	metrics on $\mathbb{R}^p$ with $N = (A^\mathrm{T}A)^{-1}$
N_ℓ	$\mathbb{R}^{p_\ell\times p_\ell}$	metrics on $\mathbb{R}^{p_\ell}$ with $N_\ell = (A_\ell^\mathrm{T}A_\ell)^{-1}$
P	$\mathbb{R}^{q\times q}$	metrics on $\mathbb{R}^q$ with $P = (B^\mathrm{T}B)^{-1}$
p_ℓ	$\mathbb{N}$	number of columns of A_ℓ
$\mathcal{P}_\ell$	$\mathbb{R}^{n\times n}$	projector on span A_ℓ in $\mathbb{R}^n$
q	$\mathbb{N}$	number of columns of B
R	$\mathbb{R}^{p\times q}$	for calculation; $R = MTQ$
$R_{\ell\ell'}$	$\mathbb{R}^{p_\ell\times p_{\ell'}}$	for calculation; $R_{\ell\ell'} = M_\ell T_{\ell\ell'} M_{\ell'}$
s	$\mathbb{N}$	number of columns of S; $s = \sum_\ell p_\ell$
S	$\mathbb{R}^{n\times(p+q)}$	a matrix on which to run a PCA, $S = CD^{1/2}$
	$\mathbb{R}^{n\times s}$	a matrix on which to run a PCA, $S = CD^{1/2}$
T	$\mathbb{R}^{p\times q}$	for calculation; $T = A^\mathrm{T}B$
$T_{\ell\ell'}$	$\mathbb{R}^{p_\ell\times p'_\ell}$	for calculation; $T_{\ell\ell'} = A_\ell^\mathrm{T}A_{\ell'}$
v	$\mathbb{R}^{p+q}$	principal axis of S, $S^\mathrm{T}Sv = \lambda v$
v	$\mathbb{R}^s$	principal axis of S, $S^\mathrm{T}Sv = \lambda v$
y	$\mathbb{R}^n$	principal component of S, $y = Sv$

Notations of Chapter 9

Symbol	in space	Meaning
d		a distance in a discrete metric space
$d(i, j), d_{ij}$	$\mathbb{R}^+$	distance between points i and j in M
D	$\mathbb{R}^{n \times n}$	matrix of pairwise distances in a discrete metric space
G	$\mathbb{R}^{n \times n}$	Gram matrix
g_{ij}	$\mathbb{R}$	a term of the Gram matrix
$i, j, \ldots$	M	points in M
Λ	$\mathbb{R}^{n \times n}$	diagonal matrix of singular values of G
M		a discrete metric space
n	$\mathbb{N}$	cardinality of M: $n = |M|$
r	$\mathbb{N}$	rank of the MDS
Σ	$\mathbb{R}^{n \times n}$	square root of singular values of G: $\Sigma = \Lambda^{1/2}$
U	$\mathbb{R}^{n \times n}$	matrix in SVD of G: $G = U \Lambda U^{\mathsf{T}}$
x_i	$\mathbb{R}^r$	point associated to i through the MDS
X	$\mathbb{R}^{n \times r}$	matrix with rows being the points in $\mathbb{R}^r$ associated to M through the MDS
$\mathcal{X}$	$\mathbb{R}^{n \times r}$	point cloud associated to M through the MDS

Chapter 1
Introduction

Discovering patterns and rules from data has been a major driver in scientific discoveries, highlighted by Kepler's work about the rules governing the movements of planets. In many fields, progress often comes from a fruitful cross-fertilization between an empirical approach based on data and a theoretical approach based on the consequences of a few accepted fundamental principles. Myriads of quantitative data are currently available in many domains of application, like Astronomy, Earth Science, and Genomics, to name but a few. Data are often multidimensional, and algorithms for processing them are often polynomial in time with the dimension of the data (like computing an SVD of an $n \times n$ matrix which is cubic with n). This makes some algorithms nonoperational for large dimensions and has led to the *curse of dimensionality*, coined by Bellman [Bel57].

Let us consider some data living in a high-dimensional space. As the complexity of their analysis often comes from their high dimensionality, an aim of data analysis is to build a summary of them in a low-dimensional space, through a projection, while preserving as much of the information contained in their structure as possible. Such a summary is called here a *meaningful summary*. There are at least two motivations for this. First, the structure of the dataset may be blurred by noise, unseen at first glance, and one objective is to filter out the noise to characterize the structure as a meaningful summary. Second, some computations that are unfeasible in spaces of large dimension because of the *curse of dimensionality* may become feasible on a meaningful summary in lower dimension. We present in this book a diversity of settings where a meaningful summary can be built by linear projection onto low-dimensional affine spaces.

Let $\mathbf{x} = (x_1, \ldots, x_N)$ with N large represent N i.i.d. realizations of a random variables. The law of large numbers tells us that the empirical mean converges almost surely to the population mean. This is a *blessing of dimensionality*: The result holds for high-dimensional spaces. Another example is classical: Let $\mathbb{S}(n, r)$ be the ball of radius r in $\mathbb{R}^n$: $\mathbb{S}(n, r) = \{x \in \mathbb{R}^n \mid \|x\| \leq r\}$. Let us uniformly distribute m points in $\mathbb{S}(n, r)$. Then, if n is (very) large, almost all points are located

© The Author(s), under exclusive license to Springer Nature Switzerland AG 2025
A. Franc, *Linear Dimensionality Reduction*, Lecture Notes in Statistics 228,
https://doi.org/10.1007/978-3-031-95785-7_1

on the surface of the ball: The distribution of $\|x\|$ is peaked on r. Moreover, if we look closely at the skin, most of the points are located on the equator (whatever the perspective, i.e., the choice for the North Pole). These types of results are well understood as *measure concentration phenomena*, which are a blessing of high dimensionality (see [Led01, BLM13]). It happens that such a blessing exists for linear dimensionality reduction: Large point clouds living in spaces of large dimension can be projected with minimal distortion on a space of (much) lower dimension. If the dimension of the space is N, the dimension of the target space which respects distances is in Log N. This is the Johnson-Lindenstrauss lemma (see Annex B.5 for details). As (this is the "essence" of PCA) there is a connection between low rank approximation of matrices and projections of point clouds on spaces of low dimension, this leads to accurate best low rank approximations of matrices with algorithms, called *random projection*, which can be implemented in large dimension (see [Vem04, HMT11]). This book is organized as presenting such an approach, i.e., implementing the best low rank approximation of large matrices using random projection for most classical settings of linear dimensionality reduction.

1.1   Multivariate Data Analysis and Pattern Discovery

When data are multidimensional, their analysis is classically referred to as Multivariate Data Analysis (MDA). Statistics is about inference of some models from data, and MDA is about inferring some patterns from data, like correlation structure between items and features, or individuals and variables. As such, it is now part of Data Mining, Statistical Learning, and Machine Learning. This can be summarized by the subtitle of [HTF09]: data mining, inference, and prediction. Tibshirani has provided a dictionary between statistics and machine learning, referred to in [Mur12], partly given here:

Machine learning	Statistics
Weights	Parameters
Learning	Fitting
Supervised learning	Regression/classification
Unsupervised learning	Density estimation, clustering

Beyond this dictionary, there is genuine innovation in machine learning, which is about inferring meaningful patterns in data in a very elaborate way, far beyond data analysis (see e.g. [Mur12, SSBD14]). A grail is to mimic the way a brain learns from experience. This is beyond the scope of this book.

MDA is a tool for learning patterns about correlations between features and items, which is one type of data structure among many others. Supervised or unsupervised learning may rely, at one step or another, on techniques inherited from multivariate data analysis, in a same way that multivariate data analysis relies at one step or another on tools inherited from linear algebra. This can be formalized by the following succession of steps:

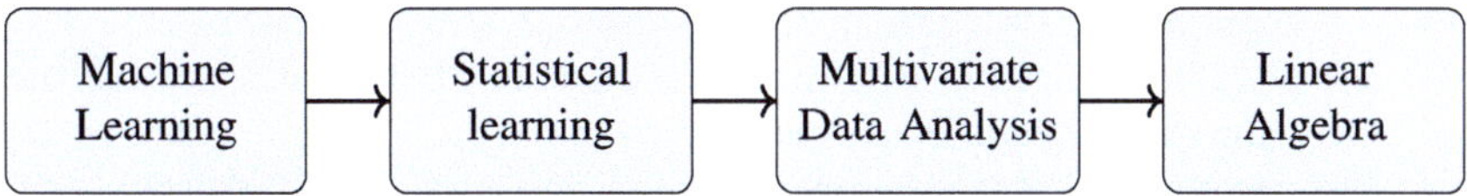

where an arrow from box A to box B means that what is in box B represents a subset of what is in box A. Another way to present it is to say that one way to solve a problem in box A is to call for a domain contained in box B. We have this inheritance from left to right providing solutions in mind for this book, with an objective of implementation of calculations for very large datasets.

There is a second reason for emphasizing the role of MDA in Machine Learning. Murphy recognized two approaches in Machine Learning [Mur12, sec. 1.1.2], which are consistent with the synthetic table of Tibshirani:

- A predictive or supervised approach, where a response variable is predicted from some features, from a set of observations where the response is known, called the training set
- A descriptive or unsupervised approach, where the objective is to find some interesting patterns, which is referred to as *pattern discovery* or *pattern recognition*.

Many of these techniques work well in low dimension, i.e., when the number of features to work with is small. MDA plays a key role in such a framework to produce a meaningful low-dimensional representation of an array connecting items and features, as close as possible to the original array. This is called Dimensionality Reduction (see, e.g., [LV07]), and one of the key (linear) techniques is therefore Principal Component Analysis (PCA, see Chap. 2). So, one can say that MDA paves the way for elaborate machine learning processes.

Notes and References   There exist many excellent surveys for learning patterns from data. See, e.g., [CST00, Bis06, Mur12, SSBD14].

1.2 Multivariate Data Analysis and Linear Algebra

Among the variety of approaches and models available for inferring patterns from data, this book focuses on linear dimensionality reduction, both as a goal and as a first step to make more elaborate and diverse algorithms feasible, as explained above. One reason for this is that it is based on a classical association between linear algebra, geometry in Euclidean spaces, and linear models in statistics, which are well-known and well-understood domains, where solutions to a problem are often known in closed form. Hence, linear dimensionality reduction is at the crossroads of three domains:

→ It is about finding structures in data presented as arrays. This is algebra.
→ It is possible to attach to an array a point cloud in a Euclidean space and study the shape of the cloud. This is geometry.
→ Data are modeled as realizations of random variables. This is statistics.

Hence, MDA is at the crossroads between algebra, geometry, and statistics.

As arrays of data are matrices, MDA heavily relies on Linear Algebra. Producing a matrix A is one of the most classical mathematical formalizations of some information gathered on a set of items. Let us consider a set of n items each characterized by p variables. The rows $i \in [\![1, n]\!]$ are the items, and the columns $j \in [\![1, p]\!]$ are some variables, often referred to as features in machine learning. The value of the feature j for the item i is the coefficient a_{ij} of the matrix. One objective is to describe how items and features are related, which is usually addressed by a low rank approximation of the matrix A.

A point cloud is a geometric object associated with such a feature matrix. Provided the variables are quantitative, i.e., numbers in $\mathbb{R}$, a set of n points in $\mathbb{R}^p$ is built from A, with one point $a_i = (a_{i1}, \ldots, a_{ip}) \in \mathbb{R}^p$ for item i. This point cloud as a geometric object is denoted $\mathcal{A}$. In many real cases, the points are located in a low-dimensional manifold. Finding such a manifold is called *dimensionality reduction*. Efficient and well-understood techniques exist when the manifold is linear (or affine): The best approximation of $\mathcal{A}$ by a projection in a space of dimension r can be solved by finding the best approximation of A by a matrix of rank r.

Row i of matrix A or point $a_i \in \mathcal{A}$ can be modeled as the realization of a set of random variables $(X_1, \ldots, X_p)$. Questions of interest include the study of the dependence structure of the X_j, given by the variance-covariance matrix of observed data, and exhibiting low-dimensional latent variables z, such that observations x can be modeled as $\mathbb{P}(x \mid z)$.

Many of these techniques have been progressively selected as core techniques over several decades since the beginning of the twentieth century. They are often presented and studied within the algebraic framework of linear algebra. For example, Principal Component Analysis (PCA, an iconic method in linear dimensionality reduction) can be built as a consequence of the Singular Value Decomposition of

A: $A = U \Sigma V^{\mathrm{T}}$. These methods are constantly coevolving with numerical linear algebra and have benefited from[1] two revolutions:

The computing revolution: The development of computing infrastructures especially in the 1970s has led to the mushrooming of scientific libraries first for mainframes and later for a range of machines from laptops to computing cluster.

The massive data revolution: Many sensors now produce myriads of bytes of data, like telescopes, satellites, sequencers, etc., raising the challenge to overcome the walls of time and memory while implementing those methods on massive datasets, leading to matrices of very large dimensions (like 10^5 to 10^6 rows or columns).

The progress in these domains is due simultaneously to the derivation of new algorithms (like the random projection method for computing the Singular Value Decomposition of a large dense matrix, see [HMT11]), development of new paradigms for implementing some algorithms (like Message Passing Interface for distributed-memory parallelization), and progress in the technology of computing infrastructures (like Graphic Processing Units).

Existing methods in linear dimensionality reduction, illustrated in Sect. 1.3, can be organized into a small set of iconic methods, knowing that such a classification is far from being either unique or universally adopted. However, most textbooks presenting these methods progressively reach an agreement on the poles around which to organize their variety and similarities/dissimilarities. We will not discuss this here but select among many possibilities a small set of methods, organized along a variety of questions they answer. These methods are presented in the following table.

Acronym	Method	Section
PCA	Principal Component Analysis	2
PCAsc	Scaled-Centered PCA	2.7
PCAda	PCA with Double Averaging	2.7
PCAmet	PCA with Metrics	4
CoA	Correspondence Analysis	5
PCAiv	PCA with Instrumental Variables	6
CCA	Canonical Correlation Analysis	7
MCoA	Multiple Correspondence Analysis	8
MDS	Multidimensional Scaling	9

---

[1] at least....

Key Observation Every method, as will be shown in the following chapters, can be organized as a sequence of pretreatment—SVD and posttreatment, as in the following pipeline of calculations:

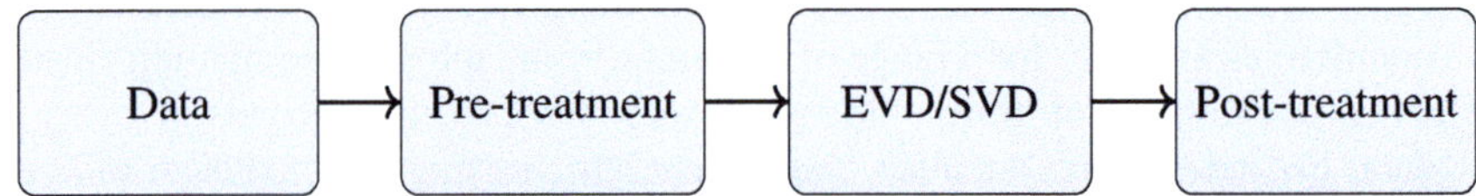

Then, as far as scaling the dimension of the data arrays is concerned, the possibilities offered by the scaling of SVD with random projection can be transferred to any method organized as such. This is what this book aims to present.

Notes and References Multivariate Data Analysis is a classical domain in data analysis, still relevant, and underestimated (although [Jol02] and a current search for "Principal Component Analysis" on Google Scholar provided more than 3 million hits). It has been developed along several lines over the twentieth century. Most of the seminal papers were published before 1935. Two trends coexist: a statistics-oriented trend, developed by the Anglo-Saxon school, in UK, USA, India, and Scandinavia, and the French school, more algebraic and geometrical, developed in the 1960s under the leadership of J.-P. Benzecri. A comprehensive and very informative paper that compares both approaches with historical insight is [TY85]. Pearson's seminal 1901 paper on PCA however is geometrical, and Hotelling's presentation 30 years later is algebraic. Several recent and excellent textbooks exist for a global presentation of MDA or dimensionality reduction, like [And58, MKB79, CC80, LV07, Ize08, Wan12]. Each has a special flavor: [And58] is the seminal book on MDA (Anderson was in Stanford) and has been a bedside book of statisticians for decades; [MKB79] is the most comprehensive, with all demonstrations of results presented as theorems, [CC80] establishes a link with statistics and is easier to read, [LV07] goes beyond linear methods, focusing on nonlinear methods, [Ize08] is comprehensive too, focusing on a diversity of examples, and [Wan12] explicitly addressed the new challenge raised by massive data and work in high-dimensional spaces. The seminal books for French school, more geometrical, are [Ben73b, Ben73a], and [CP79] who introduced the duality diagram as a unifying framework (for presentation of the duality diagram, see also [DD07, DlCH11]). See [LMT77, LMF82, LMP00, EP90] among others. The very nice paper [PCY79] gives a historical sketch of the French school and a comparison with the Anglo-Saxon school.

1.3   A Diversity of Classical Settings

Even if mushrooming today, Multivariate Data Analysis is a domain which has decades of research and practice in many domains of application, including ecology, genetics and genomics, sociology, econometrics, etc. Linear algebra is behind the scenes, reducing the main recognized methods to a single decomposition of a matrix: SVD, as mentioned above. Selecting one method or another, according to the dataset and the question asked, may be driven by specifying some pretreatments and posttreatments. Taking this route requires a slight effort in abstraction (e.g., see a data array as a matrix, on which some computations can be done). We present here the more classical settings with concrete illustrations, especially in ecology (one may say that data analysis coevolved with ecology in the 1970s).

As a simple example, let us imagine a set of sites and on each site a list of registered environmental variables, such as temperature and precipitation for weather, pH, and other indicators like C/N ratio for soils, etc. The data can be presented as an array named A, with sites as rows (say, i) and variables as columns (say, j). Elements are named a_{ij}, with $1 \leq i \leq n$ (n sites) and $1 \leq j \leq p$ (p variables), with a_{ij} being the value of variable j for site i. In this setting, rows represent items, and columns represent variables or features. Several questions can be asked and answered. One of them which will occupy us for the rest of the book is to take into account the correlation structure between variables to summarize the array while keeping much of the information it contains. Such an approach is called *Principal Component Analysis*, denoted *PCA*, and is presented in Chap. 2.

The correlation structure between variables can be formalized by equipping the space of columns, i.e., $\mathbb{R}^n$, with a Euclidean structure, hence an inner product, denoted $\langle ., . \rangle$, and a distance. Classical PCA relies on the standard Euclidean structure, i.e., $\langle x, y \rangle = \sum_{i=1}^{n} x_i y_i$ if $x = (x_1, \ldots, x_n)$ and $y = (y_1, \ldots, y_n)$ both belong to $\mathbb{R}^n$. The calculation of the principal axis and component for a given array (which help build an accurate summary) can however be derived for any Euclidean structure. This leads to a diversity of PCAs with so-called metrics on rows and columns. Such a diversification is presented in Chap. 4.

Another common type of data array is the type contingency tables. They represent the distribution of N individuals according to the values of two qualitative or categorial variables. The rows represent one variable and the columns another. Then, a_{ij} is the number of individuals with modality i of the first variable and j of the second. One has $a_{ij} \in \mathbb{N}$ and $\sum_{i,j} a_{ij} = N$. As an example, let us imagine n sites and p plant species, and let a_{ij} be the number of plants in site i which belong to species j. The model used to summarize the array is based on assuming independence between both variables, i.e., $\mathbb{P}(I = i, J = j) = \mathbb{P}(I = i) \times \mathbb{P}(J = j)$, where $\mathbb{P}(X = x)$ is the probability that random variable X has value x (here, I is the first variable and J the second). If the model is exact, it suffices to know $a_{i+} = \sum_j a_{ij}$ for each i and $a_{+j} = \sum_i a_{ij}$ for each j to reconstruct the contingency table: $a_{ij} = \frac{1}{N} a_{i+} a_{+j}$. This (after some technicalities to deal with more than one axis) is called *Correspondence Analysis*, denoted *CoA*, and it will be shown that it can be written as a PCA with metrics via a judicious choice of metrics, i.e., the way to quantify the distances between rows and columns. This is presented in Chap. 5.

Here is another setting for *CoA*. The array $A = (a_{ij})_{i,j}$ is classically simplified as a presence/absence array, where $a_{ij} \in \{0, 1\}$ and represents the absence/presence of species j in site i. There is a classical observation, which can be explained rigorously, see Chap. 8, that even if in such a case A is no longer a contingency table, it can be analyzed with the same method, i.e., Correspondence Analysis. This suggests that these different methods are linked, which has been known since the 1970s. These links are the guiding thread of this book.

The next setting involves two arrays or more, with the same items, but different sets of variables or features. It is presented here as a setting with an array A being the contingency table of plants distributed among sites (or presence/absence) and a second array B being the array of the values of environmental variables for each site. Three types of questions can be asked, and each leads to a specific method.

The first natural question is to ask whether A and B, with the same items but different features, yield the same organization of items from the correlation structure of features. As we are working with linear dimensionality reduction, this involves finding two new features as a linear combination of the columns of A on one side and of B on the other, with maximal correlation. This is *Canonical Correlation Analysis* and is presented in Chap. 7. Now, in ecology, it is known that environmental variables (the columns of B) drive the presence/absence of plant species (the columns of A). This involves explaining the columns of A by the columns of B and can be implemented by finding the best linear regression with the columns of B as regressors common for all columns of A. This is PCA-IV, or PCA with Instrumental Variables, and is presented in Chap. 6. Let us go one step further. It is not always possible in daily basis field work to measure all relevant environmental variables. A common practice is to develop bioindication, i.e., infer the environmental variables site by site from the presence/absence of plant species. This can be formalized as a PCA-IV as well, but with columns of A (the regressors) as instrumental variables for building a best approximation of B.

For data arrays, such approaches mimic the correlation between variables: If $\mathbf{y}$ and $\mathbf{y}'$ are two variables on the same set of items, one can (i) define the correlation ρ between them, which is extended to the correlation between arrays by *CCA* and (ii) infer the best regression of $\mathbf{y}$ by $\mathbf{y}'$ or of $\mathbf{y}'$ by $\mathbf{y}$, which is extended to arrays by *PCA-IV*. Let us note that, from a historical perspective, regression of $\mathbf{y}$ by $\mathbf{y}'$ is different from regression from $\mathbf{y}'$ by $\mathbf{y}$, i.e., $\mathbf{y}$ and $\mathbf{y}'$ do not play the same role. In his pioneering work, Pearson was motivated to find an approach such that $\mathbf{y}$ and $\mathbf{y}'$ play a symmetric role.

One can ask legitimately why limit the correlation structure between two arrays and not between three or more? An immediate answer is that there is no canonical definition of a correlation between three variables. It follows that there is no canonical definition of a correlation between the three principal axes of three arrays or more. This difficulty has been solved by establishing a correspondence between Correspondence Analysis and Canonical Correlation Analysis between two arrays. Such an approach can naturally be extended to three or more arrays, leading to *Multiple Correlation Analysis*, presented in Chap. 8.

Finally, let us recall that it is possible to associate a point cloud $\mathcal{A}$ with a dataset $A \in \mathbb{R}^{n \times p}$: A point $a_i \in \mathbb{R}^p$ is associated with row i of A as the point with coordinates $(a_{ij})_j$. A point cloud can be given by the list of coordinates of its points or, up to an isometry, as the set of pairwise distances between its points, i.e., the array of $d(i, j) = \|a_i - a_j\|$ in a Euclidean space. It is easy to compute pairwise distances knowing the points but is less trivial to find the points knowing their distances (up to an isometry). Such an approach called *Multidimensional Scaling* is subtle even if well understood and presented in Chap. 9.

1.4 Some Issues That Are Not Addressed in This Book But Are Related

This book is only the tip of a much richer iceberg. It is restricted to the search for patterns in data described by low-dimensional linear or affine varieties. There are two reasons for this: Firstly, there are well-understood algorithms of reasonable complexity for constructing a solution in closed form, but more importantly, to the best of our knowledge it is hopeless to look for an algorithm that has a dataset as input and identifies a shape of any kind as output. One reason is that the notion of shape is vague, even if very evocative.

Currently there are (at least...) three ways in which dimensionality reduction can be developed:

◇ Stay in the domain of linear dimension reduction, but work with distances not associated with a Euclidean structure, like L^1 distances. This is a very difficult domain still in progress.
◇ For the geometric setting, look for low-dimensional nonlinear manifolds which carry most of the information on the structure of the dataset. This is known as *manifold learning*. A survey of the most important methods can be found, for example, in [LV07] and a contemporary accessible survey in [MZ24].
◇ Some extensions are specific to extending MDS, like Sammon's mapping, ISOMAP, etc.

Much information about these trends can be recovered from Wikipedia, starting, for example, with the article https://en.wikipedia.org/wiki/Nonlinear_dimensionality_reduction.

1.5 Some Examples of Software Implementation

There exist many libraries in different languages for implementing these linear methods. Here, we focus on R and `python` because they are each a standard in data analysis, although we present a C++ library too. We briefly present `ade4`, `FactoMineR`, `scikit-learn`, and `diodon` as recommendations, with links to their websites.

R For R, see https://cran.r-project.org/. There exist many packages providing the possibility to run one or the other method. PCA, for example, is part of

MASS. The package `ca` (see https://cran.r-project.org/web/packages/ca/index.html) is devoted to `CoA` and `MCoA`. There is no so-called *task view* on Multivariate Data Analysis, but one called *environmetrics* on Analysis of Ecological and Environmental Data, available at https://cran.r-project.org/web/views/Environmetrics.html. Two packages provide a diversity of methods on MDA: `factoMineR` and `ade4`. `FarctoMineR` is devoted to three methods: `PCA`, `CoA`, and `MCA`, being very faithful to the approach by the so-called French School. It is available at http://factominer.free.fr/index.html. There is a companion book [HLP17]. `ade4` has been developed for several decades by the same team in Lyon and proposes a thorough view on a diversity of methods in MDA applied to community ecology. See their site at https://adeverse.github.io/ade4/. It is accompanied by a book, [TDD$^+$18], which both explains the methods and develops some examples. Their guiding thread for encompassing all methods in the same framework is the duality diagram, which is an alternative to SVD. But it does not provide a link to scaling with random projection.

Scikit-learn `scikit-learn` is a versatile library for statistical learning written in `python` and based on `numpy`, `scipy`, `matplotlib`. See https://scikit-learn.org/stable/. There is in the front page a link toward a set of methods for dimensionality reduction, at https://scikit-learn.org/stable/modules/decomposition.html. When compared to the methods presented in this book, it offers more possibilities to PCA and closely related methods (like incremental PCA, sparse PCA, kernel PCA, etc.). It offers the possibility to work with randomized SVD. `CoA` however is currently not available. It provides however CCA (as `sklearn.cross_decomposition.CCA`) and PLS which is known to be equivalent to `PCAiv`.

Diodon Some of the algorithms presented in these notes including the integration of Randomized SVD have been implemented in:

⋄ A C++ library, called `cppdiodon`, publicly available at https://gitlab.inria.fr/diodon/cppdiodon, which has been published in HAL and Software Heritage (see https://hal.science/hal-04971369)
⋄ A Python library, built on `numpy` and `matplotlib`, called `pydiodon`, publicly available at https://gitlab.inria.fr/diodon/pydiodon, which has been published in HAL and Software Heritage (see https://hal.science/hal-05021182)

The reason why we propose the C++ library is that it really is optimized for very large datasets, by associating randomized SVD, distributed memory, and task-based programming with `Chameleon`. We could perform an MDS on a $10^6 \times 10^6$ dense dataset (distance array) in 500 seconds, including reading the data. If randomized SVD is efficiently available in `pydiodon`, neither distributed memory nor task-based programming is available in the `python` version, for which datasets up to a size of $10^5 \times 10^5$ can be processed using SVD with random projection.

Chapter 2
Principal Component Analysis (PCA)

Abstract This chapter presents the core method around which all the others will be organized, as well as its main variations: Principal Component Analysis (PCA). This method is the backbone of the book. The three main domains in which it can be developed are presented: algebra, with the approximation of a given matrix by a low rank matrix; Euclidean geometry, which is more intuitive and involves projecting a point cloud on a low-dimensional affine space; and statistics, where it involves exhibiting a correlation structure between variables. We recall that performing a PCA on a matrix A is equivalent to calculating its Singular Value Decomposition (SVD). This allows us to inherit all the advances made in the calculation of an SVD in high dimension, in particular by random projection as presented here, in the calculation of a PCA. We also introduce the main concepts (principal axis and principal coordinate) and the classical evaluations of the quality of the approximation which will be used throughout the book, i.e., by the variety of methods presented in the introduction.

PCA is a dimensionality reduction technique which can be read in three different ways:

Geometric: A point cloud $\mathcal{A}$ of n points in $\mathbb{R}^p$ being given, as well as an integer $r < p$, find an affine subspace $E \subset \mathbb{R}^p$ of dimension r such that the projection $\mathcal{A}_E$ of $\mathcal{A}$ on E is as close as possible to $\mathcal{A}$.

Algebraic: A $n \times p$ matrix A being given, as well as an integer $r < p$, find a matrix A_r of rank r such that $\|A - A_r\|$ is minimum with Frobenius (ℓ^2) norm.

Statistical: A set of p random variables being observed on n items independently, find r independent linear combination of these variables with maximum variance (the first one is with maximum variance, the second one is uncorrelated with the first one and with maximum variance, and so on).

Here, we adopt the algebraic viewpoint but start with giving some links with the geometric one, which is important for visualization of point clouds. Geometric approach is about dimensionality reduction, and algebraic approach is about the best low rank approximation. The best low rank approximation has many applications in numerical linear algebra, e.g., for optimizing matrix $\times$ matrix product. There

A. Franc, *Linear Dimensionality Reduction*, Lecture Notes in Statistics 228,
https://doi.org/10.1007/978-3-031-95785-7_2

are many links between algebraic and statistical approach too, which are merely sketched here.

Notes and References The historical development of PCA is well known and well documented. According to [Bas94, chap. 3], its origin can be traced in the work of Bravais in 1846 (where the notion of principal axis emerged, in "in the form of rotating an ellipse to *axes principaux* in order to achieve independence in a multivariate normal distribution"). Classically, its origin is attributed to Pearson in [Pea01] under the guise of both a statistical and a geometrical derivation, as an extension of the linear regression. The aim is statistical, but the idea behind is clearly geometric. Pearson's aim was to escape from the non-symmetry of dependent and independent variables in linear regression (the regression line changes if the status dependent versus independent are exchanged between variables), by giving equal status for variations on both types of variables. He was led to find the best fit of, say, a system of points in the plane by a line. One guise of his approach is statistical, in the sense that it uses the notions of mean, variance, and standard deviation of a sample, but the notion of statistical model as it is understood nowadays was not available at this time. So, Pearson's approach can be qualified as geometrical. The term Factor Analysis has been introduced by Thurstone in 1931 (Thurstone, L., 1931, Multiple Factor Analysis, *Psychological Review*, **38**:406–427). His purpose was to find a general method for finding factors which could explain correlations, following an idea published by Spearman in 1904. The presence of an underlying model in factor analysis has been the cause of numerous and fierce discussions (see [Jol02]). Hotelling gave an algebraic framework in 1933 (in Hotelling, H., 1933, Analysis of a complex of statistical variables into principal components. *J. Educ. Psychol.*, **24**:498–520), where the term PCA first appeared. Following the work of Pearson, he showed that the principal axes are the eigenvectors of the covariance matrix of the sample. PCA is based on SVD of a matrix. The link between PCA and SVD is classically attributed to a theorem published by Eckart and Young in 1936 [EY36]. Classical textbooks in the Anglo-Saxon literature dedicated to PCA are [Jac91] and [Jol02]. The triple nature of PCA (algebraic, geometrical, and statistical) can be found in [VMS16], where chapter 2 is a thorough presentation of PCA. PCA is developed in every textbook in Multivariate Data Analysis, like [MKB79, CC80, Ize08]. [Wol87] is a survey of PCA tools with an introduction on main milestones in the development of the method. Classical textbooks in French literature are [CP79, LMF82, LMP00]. A recent survey of PCA and its recent developments can be found in [JC16].

PCA is probably one of the most used tools in multivariate statistics. It is known under various names according to the field where it is applied, as mentioned in https://en.wikipedia.org/wiki/Principal_component_analysis: Principal Component Analysis, Karhunen–Loève transform (KLT) in signal processing, proper orthogonal decomposition (POD) in mechanical engineering, and empirical orthogonal functions (EOFs) in meteorological science, among others.

If the theory can be derived for any pair (p, r) of dimensions with $1 \leq r < p$, it is most interesting when p is large and $r \ll p$.

2.1 Setting the Problem

Let us recall that the norm here and in the following sections is the Frobenius or ℓ^2 norm unless otherwise stated:

$$\|x\| = \left(\sum_i x_i^2 \right)^{1/2} \tag{2.1.1}$$

for a vector and

$$\|A\| = \left(\sum_{i,j} a_{ij}^2 \right)^{1/2} \tag{2.1.2}$$

for a matrix.

The problem can be set as follows:

- **Algebraic approach**

$$
\begin{aligned}
&\text{Given} \quad && A \in \mathbb{R}^{n \times p} \\
&&& 0 < r < p \\
&\text{Find} \quad && A_r \in \mathbb{R}^{n \times p} \\
&\text{with} \quad && \text{rank } A_r = r \\
&\text{such that} \quad && \|A - A_r\| \quad \text{minimal}
\end{aligned}
$$

- **Geometric approach**

In this approach, a cloud denoted $\mathcal{A}$ of n points in $\mathbb{R}^p$, labeled $(a_i)_i$ with $1 \leq i \leq n$, is associated with a matrix $A \in \mathbb{R}^{n \times p}$, where the coordinates of the point a_i are the row i of A, denoted a_i as well. The point cloud associated with A_r is denoted $\mathcal{A}_r$.

> Given a point cloud $\mathcal{A} = (a_1, \ldots, a_n)$
> $a_i \in \mathbb{R}^p$
> $0 < r < p$
>
> Find   a subspace $E_r \subset \mathbb{R}^p$
> with $\dim E_r = r$
>
> such that $d(\mathcal{A}, \mathcal{A}_r)$ minimal
>
> where $d(\mathcal{A}, \mathcal{A}_r) = \sum_i \|a_i - \widetilde{a}_i\|^2$
> and $\widetilde{a}_i$ is the projection of a_i on E_r

In most of the cases, the point cloud is analyzed after a translation to its barycenter defined by $g = \frac{1}{n} \sum_{i=1}^{n} a_i$. Otherwise, the first axis represents simply the barycenter and not the inner structure of the point cloud. E_r is an affine space which contains the origin

2.2   Solving the Problem

The solution to this problem is well known (see any textbook mentioned in the introduction of this section). Finding the subspace E_r is finding an orthonormal basis for it. Let us fix a subspace $E_r \subset \mathbb{R}^p$ of dimension r. If $\widetilde{a}_i$ is the projection of a_i on E_r, we have, by Pythagoras' theorem,

$$\forall\, i \in [\![1, n]\!], \quad \|a_i\|^2 = \|\widetilde{a}_i\|^2 + \|a_i - \widetilde{a}_i\|^2.$$

Then, setting $\sum_i \|a_i - \widetilde{a}_i\|^2$ minimum is equivalent to setting $\sum_i \|\widetilde{a}_i\|^2$ maximum. We then can set PCA as

$$
\begin{aligned}
&\text{Given a point cloud } \mathcal{A} = (a_1, \ldots, a_n) \\
&\qquad\qquad\qquad a_i \in \mathbb{R}^p \\
&\qquad\qquad\qquad 0 < r < p \\[2ex]
&\text{Find} \qquad\quad\ \text{a subspace } E_r \subset \mathbb{R}^p \\
&\text{with} \qquad\quad\ \dim E_r = r \\[2ex]
&\text{such that} \qquad \sum_i \|\widetilde{a}_i\|^2 \quad \text{maximal} \\[2ex]
&\text{where} \qquad\quad \widetilde{a}_i \text{ is the projection of } a_i \text{ on } E_r.
\end{aligned}
\tag{2.2.1}
$$

Let us consider the simple case $r = 1$ and $0 \in E_r$. If u with $\|u\| = 1$ is a basis of E_1, we have $\widetilde{a}_i = \langle a_i, u \rangle u$, and the optimization problem can be stated as

$$
\left|
\begin{array}{lll}
\text{Given a point cloud } \mathcal{A} = (a_1, \ldots, a_n) & & \\
\qquad\qquad\qquad a_i \in \mathbb{R}^p & & \\
& & \\
\text{Find} & \text{a vector } u \in \mathbb{R}^p & \\
\text{with} & \|u\| = 1 & \\
& & \\
\text{such that} & \|Au\| \quad \text{maximal.} &
\end{array}
\right. \tag{2.2.2}
$$

Indeed, $\widetilde{a}_i = \langle a_i, u \rangle u$. Then $\sum_i \|\widetilde{a}_i\|^2 = \sum_i \langle a_i, u \rangle^2 = \|Au\|^2$. The elements $\langle a_i, u \rangle$ are the coordinates of the vector $y = Au$. The solution of such a problem is classical: u is the eigenvector of $A^{\mathrm{T}} A$ associated with the largest eigenvalue

$$
A^{\mathrm{T}} A v = \lambda v, \qquad \lambda = \max\{\lambda \in \mathrm{Sp}\, A^{\mathrm{T}} A\}, \tag{2.2.3}
$$

and λ is the spectral norm of A. The Eckart-Young theorem [EY36] extends this result to $r > 1$ and states that an orthonormal basis of E_r is the set $(v_1, \ldots, v_r)$ with

$$
A^{\mathrm{T}} A v_j = \lambda_j v_j \qquad \text{with} \quad \lambda_1 \geq \ldots \geq \lambda_r \geq \lambda_{r+1} \geq \ldots \geq \lambda_p \geq 0. \tag{2.2.4}
$$

This is a direct application of the variational properties of the Rayleigh quotients (see [HJ12, sec. 4.2]). This can be written

$$
A^{\mathrm{T}} A V = V \Lambda, \tag{2.2.5}
$$

where V is the $p \times p$ matrix with column j being v_j and Λ is the diagonal matrix with terms $(\lambda_i)_i$ in the diagonal (in decreasing order). The coordinates of the point cloud $\mathcal{A}$ in new basis V are given by

$$
Y = AV, \tag{2.2.6}
$$

hence a first algorithm, for a function called PCA_EVD():

Algorithm 1 PCA of a matrix with EVD: PCA_EVD(A)

1: **input** $A \in \mathbb{R}^{n \times p}$
2: **compute** $C = A^{\mathrm{T}} A$
3: **compute** $(\lambda_\alpha, v_\alpha)$ such that $C v_\alpha = \lambda_\alpha v_\alpha$, or $CV = V \Lambda$
4: **compute** $Y = AV$
5: **return** Y, Λ, V

The vectors in new basis are called *principal axes*, and the coordinates along the principal axes are called *principal components*.

It should be noted that, although this approach is the basis of all the methods presented in this text, it is very rarely applied as such: PCA is applied after the data have been preprocessed. Classical examples of preprocessing include translation of the cloud to its barycenter, known as centering, and column (variable) normalization.

2.3 Link with SVD

We present here a tight connection between PCA and SVD, which will have far-reaching consequences on numerical analysis.

Let (U, Σ, V) be the SVD of A:

$$A = U \Sigma V^{\mathrm{T}}. \tag{2.3.1}$$

Then

$$\begin{aligned} C &= A^{\mathrm{T}} A \\ &= (V \Sigma U^{\mathrm{T}})(U \Sigma V^{\mathrm{T}}) \\ &= V \Sigma^2 V^{\mathrm{T}} \qquad \text{as} \quad U^{\mathrm{T}} U = \mathbb{I}_n, \end{aligned} \tag{2.3.2}$$

and

$$CV = V \Sigma^2 \qquad \text{as} \quad V^{\mathrm{T}} V = \mathbb{I}_p. \tag{2.3.3}$$

Hence, V as new basis for PCA of A is the matrix V in SVD of A, and $\Lambda = \Sigma^2$. We have

$$\begin{aligned} Y &= AV \\ &= U \Sigma V^{\mathrm{T}} V \\ &= U \Sigma. \end{aligned} \tag{2.3.4}$$

This yields a second algorithm for PCA:

Algorithm 2 PCA of a matrix with SVD: PCA_SVD(A))

1: **input** $A \in \mathbb{R}^{n \times p}$
2: **compute** $U, \Sigma, V = \text{SVD}(A)$
3: **compute** $\Lambda = \Sigma^2$
4: **compute** $Y = U \Sigma$
5: **return** Y, Λ, V

About the Choice Between SVD and EVD Adopting the SVD viewpoint has some numerical advantages:

→ There exist efficient and stable algorithms for computing an SVD.
→ It avoids a matrix × matrix computation ($C = A^{\mathsf{T}}A$) which can be costly when n and p are large.
→ It leads to easier generalization with instrumental variables or metrics on row or column space (see Chaps. 4 and 6).
→ If the dimensions n and p are (very) large, the SVD can be computed with random projection (see [HMT11]). Such a calculation is presented below.

However, matrix $C' = \frac{1}{n} A^{\mathsf{T}}A$ is the variance-covariance matrix of the distribution of the variables in the statistical approach and must not be ignored.

Summary of the Results Here is a summary of the calculations:

A	$n \times p$	Data set	
C	$p \times p$	Matrix of correlations of columns of A	$C = A^{\mathsf{T}}A$
(U, Σ, V)		SVD of A	$A = U\Sigma V^{\mathsf{T}}$
V	$p \times p$	Principal axis	$Cv = \lambda v$
Λ		Eigenvalues of C in decreasing order	$\Lambda = \Sigma^2$
Y	$n \times p$	Principal components	$Y = AV = U\Sigma$

It can be sketched as

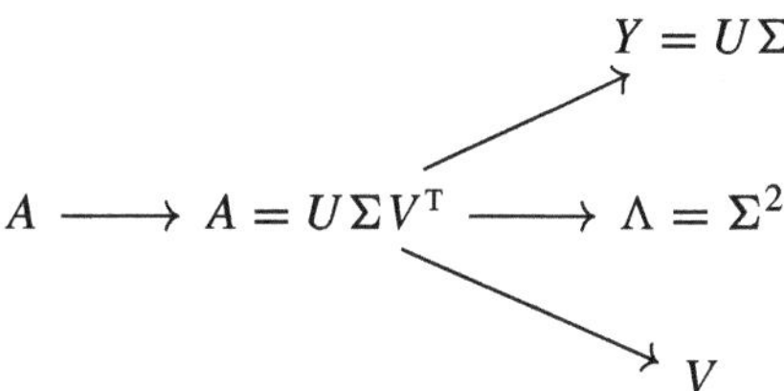

2.4 Randomized SVD

Let $A \in \mathbb{R}^{n \times p}$ with $n \geq p$. The complexity (number of operations) of the SVD of A is in $\mathcal{O}(n^2 p)$. SVD becomes untractable for large values of n and p, say, 10^4. Fortunately, there are some very efficient heuristics, with bounds on errors, to compute the first singular values and vectors, based on randomized algorithms. The idea behind is the following.

If $Q \in \mathbb{R}^{n \times k}$ is column-wise orthonormal, the projections of the columns of A on the vector space spanned by the columns of Q are the columns of

$$\widetilde{A} = Q Q^{\mathrm{T}} A.$$

If a point cloud $\mathcal{A}$ of n points in $\mathbb{R}^p$ is associated with A, the rows of $\widetilde{A}$ are the projections of the points in $\mathcal{A}$ on the k-dimensional space spanned by the columns of Q. Let us denote

$$B = Q^{\mathrm{T}} A, \qquad B \in \mathbb{R}^{k \times p}.$$

Then, $\widetilde{A} = QB$ is a rank k approximation of A ($Q \in \mathbb{R}^{n \times k}$ and $B \in \mathbb{R}^{k \times p}$). The SVD of B ($B = U_{\mathrm{B}} \Sigma V^{\mathrm{T}}$) is in $\mathcal{O}(p^2 k)$ instead of $\mathcal{O}(n^2 p)$, and we have

$$A \approx QB = Q(U_{\mathrm{B}} \Sigma V^{\mathrm{T}}) = (Q U_{\mathrm{B}}) \Sigma V^{\mathrm{T}} = U \Sigma V^{\mathrm{T}}.$$

So

$$A \approx U \Sigma V^{\mathrm{T}} \qquad \text{with} \quad U = Q U_{\mathrm{B}}. \tag{2.4.1}$$

The next step is to show that, when n and p are large, $A \approx Q Q^{\mathrm{T}} A$ with high quality for any random matrix Q (it means with probability 1 when $n \to \infty$). This comes from the Johnson-Lindenstrauss lemma which states that, for any cloud $\mathcal{X}$ of n points in $\mathbb{R}^p$, for any $\epsilon > 0$, if

$$k \geq \frac{8 \operatorname{Log} n}{\epsilon^2}, \tag{2.4.2}$$

there is a linear map

$$f : \mathbb{R}^p \longrightarrow \mathbb{R}^k,$$

which reduces the dimensionality and preserves the distance. More precisely, for any $x, y \in X$

$$(1 - \epsilon) \|x - y\|^2 \leq \|f(x) - f(y)\|^2 \leq (1 + \epsilon) \|x - y\|^2. \tag{2.4.3}$$

The bad news is that ϵ^2 is at the denominator (so k is large when ϵ is small), but the good news is that k grows with $\operatorname{Log} n$ and not n. This becomes efficient when n is large. So, Q as an orthonormal basis is built as the QR-decomposition of $Y = A\Omega$, where Ω is a random matrix (Y is in the span of A). There are several ways to choose Ω, and here we restrict ourselves to the Gaussian random projection, i.e., Ω is a random Gaussian matrix with

$$\Omega[i, j] \sim \mathcal{N}(0, 1).$$

Usually, for a good accuracy at rank k, it is advised to select Ω as $n \times k'$ with $k' = k + s$, where s is called the oversampling. Usually, taking $s = 5$ is said to be sufficient. The reader is encouraged to read [HMT11] for further details and explanations on randomized algorithms in matrix computations (what is presented here is the tip of the iceberg). The algorithm runs as follows:

Algorithm 3 SVD of a matrix with Gaussian random projection $\text{SVD\_GRP}(A, k))$

1: **input** $A \in \mathbb{R}^{n \times p}$, k as prescribed rank
2: **build** $\Omega \in \mathbb{R}^{p \times k}$, random $(\Omega[i, j] \sim \mathcal{N}(0, 1))$
3: **compute** $Y = A\Omega$
4: **compute** the QR-decomposition of Y: $Y = QR$
5: **build** $B = Q^{\mathsf{T}}A$
6: **run** the SVD of B: $B = U_{\mathrm{B}}\Sigma V^{\mathsf{T}}$,   or $(U_{\mathrm{B}}, \Sigma, V) = \text{SVD}(B)$
7: **compute** $U = QU_{\mathrm{B}}$
8: **return** U, Σ, V

Here are the dimensions of the involved matrices:

Matrix	Dimensions	Computation
A	$n \times p$	Data
Ω	$p \times k$	Gaussian random matrix
Y	$n \times k$	$Y = A\Omega$
Q	$n \times k$	$Y = QR$
B	$k \times p$	$B = Q^{\mathsf{T}}A$
U_{B}	$k \times k$	$B = U_{\mathrm{B}}\Sigma V^{\mathsf{T}}$
Σ	$k \times k$	idem
V	$k \times p$	idem
U	$n \times k$	$U = QU_{\mathrm{B}}$

and a sketch of the calculations:

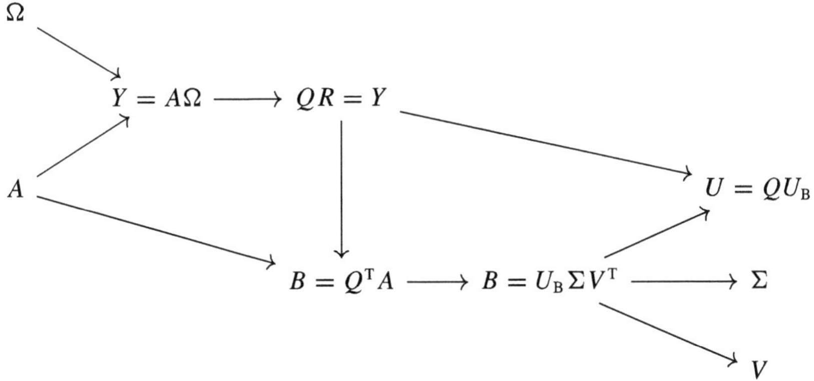

In Chap. 9, we will see that MDS relies on the SVD or EVD of the Gram matrix of a point cloud (if $\mathcal{X} = (x_i)_i$ is a point cloud with $x_i \in \mathbb{R}^p$ and $1 \leq i \leq n$, the Gram matrix G of $\mathcal{X}$ is the symmetric positive matrix of elements $g_{ij} = \langle x_i, x_j \rangle$). Then, $G = XX^{\mathrm{T}}$. MDS is solving the reverse problem: finding X knowing G. It can be solved by an SVD of the Gram matrix, which can be done by random projection when dimensions are large. However, a simple transposition leads to technicalities, which are presented and solved in Sect. 9.4.

Notes and References The property behind Random Projection (RP) is counter-intuitive: "RP, while reducing dimensionality, approximately preserves pairwise distances with high probability" [Vem04, p. 2]. This is a consequence of Johnson-Lindenstrauss lemma ([JL84], see appendix B.5 for details). [Vem04] presents a variety of domains of application of random projection. This leads to revisiting the very notion of PCA. Indeed, a point cloud and a rank being given, PCA is about finding the best subspace of dimension k as far as quality of projection is concerned. RP answer is that any randomly selected subspace of dimension k will make the job! Actually, this is not true for very small k, like first two or three axes useful for visualizing the shape of the point cloud. However, when k is significantly larger, this is true. It is possible to show (see appendix B.5) that the best choice is that k is in $\mathcal{O}(\frac{\mathrm{Log}\,n}{\epsilon^2})$, where ϵ is the relative accuracy in distance preservation. The presence of ϵ^2 at the denominator forces n to be very large for $\mathrm{Log}\,n < n\epsilon^2$. This is however true for any point cloud, including random ones. In applications, point clouds have a pattern or a structure, and RP works for much smaller values of n. See, for example, [BM01] for a comparison in accuracy and computing time with some other dimensionality reduction tools like PCA and an overview of various methods for selecting the matrix Ω (Gaussian matrix is not the only possible choice). The use of randomized algorithms to compute efficiently the SVD of a very large matrix is fully developed in [HMT11]. This section explains how SVD with Gaussian Random Projection works. To understand why it works, see Appendix B.5.

2.5   Core Algorithm for PCA

Wrapping all this together leads to a core algorithm for PCA, where the user can select which method to implement, presented hereafter in pseudocode:

Comments Here are some comments:

Why PCA_CORE?   The reason for the name is the following: PCA is a method which very seldom runs dimension reduction or approximation by a low rank matrix directly on the data matrix. Most of the times, there is a pretreatment (see Sect. 2.7), like centering and scaling column-wise, and dimension reduction is performed after pretreatment. The name PCA designates classically the whole analysis:

Algorithm 4 PCA of a matrix: PCA_CORE(A, $k = -1$, meth =SVD)

1: **input** $A \in \mathbb{R}^{n \times p}$, $k \in \mathbb{N}^* \cup \{-1\}$, meth $\in \{\text{EVD}, \text{SVD}, \text{GRP}\}$
2: **if** meth==EVD **then**
3:     **compute** $C = A^\mathsf{T} A$
4:     **compute** $(\lambda_\alpha, v_\alpha)$ such that $C v_\alpha = \lambda_\alpha v_\alpha$, or $CV = V \Lambda$
5:     **compute** $Y = AV$
6:     **if** $k > 0$ **then**
7:         $Y = Y[:, 0 : \text{k}]$; $\Lambda = \Lambda[0 : \text{k}]$; $V = V[:, 0 : \text{k}]$
8:     **end if**
9: **end if**
10: **if** meth==SVD **then**
11:     **compute** $U, \Sigma, V = \text{SVD}(A)$
12:     **compute** $\Lambda = \Sigma^2$
13:     **compute** $Y = U\Sigma$
14:     **if** $k > 0$ **then**
15:         $Y = Y[:, 0 : \text{k}]$; $\Lambda = \Lambda[0 : \text{k}]$; $V = V[:, 0 : \text{k}]$
16:     **end if**
17: **end if**
18: **if** meth==GRP **then**
19:     **compute** $U, \Sigma, V = \text{SVD\_GRP}(A, k)$
20:     **compute** $\Lambda = \Sigma^2$
21:     **compute** $Y = U\Sigma$
22: **end if**
23: **return** Y, Λ, V

---

1. A pretreatment

$$A \xrightarrow{\text{pretreatment}} A'.$$

2. The treatment itself

$$Y, \Lambda, V = \text{PCA\_CORE(A')},$$

which is denoted PCA_CORE().

Choice of the method: Three methods are described here: eigenvalues of the correlation matrix (EVD), SVD of the data matrix (SVD), and SVD with Gaussian Random Projection (GRP). Here is an advice for selecting the right method:

→   If the size of the matrix (number p of columns) is small to medium (say $\approx 10^3$), then EVD or SVD can be used.
→   If the size of the matrix is large to very large ($10^4 < p < 10^6$), Gaussian random projection must be used.

Prescribed rank: k is the prescribed rank. Its default value is $k = -1$, which means that all the eigenvalues, or singular values, components, and axes will be computed. This is relevant for methods EVD or SVD only. A rank $k > 0$ must be prescribed for method GRP. If a rank $k > 0$ is prescribed, the first k eigenvalues or singular values, components, and axis only will be computed.

2.6 Interpretation and Plotting

In this section, the geometrical viewpoint is adopted. Interpretation of the results of a PCA is about quantifying the fraction of inertia of the point cloud (i.e., variance of the associated variables or norm of the associated matrix) which is preserved by projection either on one axis or on a space spanned by r first axes. When the point cloud is centered, PCA is finding a rotation in $\mathbb{R}^p$ such that these quantities are maximal.

Let $A \in \mathbb{R}^{n \times p}$ be a matrix, $\mathcal{A}$ its associated point cloud in $\mathbb{R}^p$, and (Y, Λ, V) the PCA of A. Then, the column y_j of Y with $j \in [\![1, p]\!]$ is the vector in $\mathbb{R}^n$ of the coordinates of all the points of $\mathcal{A}$ on principal axis j. Indeed, let us have a point cloud $\mathcal{A}$ made of n points $a_i \in \mathbb{R}^p$ with a_i being row i of A. PCA of A is performing an SVD of A as

$$A = U\Sigma V^{\mathsf{T}} \tag{2.6.1}$$

and yields a new orthogonal basis $(v_1, \ldots, v_p)$ of $\mathbb{R}^p$, where v_j is the column j of V. Let us recall that

$$Y = U\Sigma. \tag{2.6.2}$$

Then

$$Y = U\Sigma = U\Sigma(V^{\mathsf{T}}V) = (U\Sigma V^{\mathsf{T}})V = AV, \tag{2.6.3}$$

which means that the rows of Y are the coordinates of the points of $\mathcal{A}$ in new basis V.

If $y \in \mathbb{R}^n$ and $v \in \mathbb{R}^p$, let us recall the notation $\otimes$ for tensor product (see also Appendix B.4)

$$y \otimes v \equiv yv^{\mathsf{T}}$$

(i.e., $(y \otimes v)_{ij} = y_i v_j$). Let us denote by y_j the column j of Y and v_j the column j of V. Then,

$$A = \sum_{j=1}^{p} y_j \otimes v_j, \tag{2.6.4}$$

from which we can derive

$$\|A\|^2 = \sum_j \|y_j\|^2. \tag{2.6.5}$$

Indeed,

$$
\begin{aligned}
\|A\|^2 &= \left\| \sum_{j=1}^{p} y_j \otimes v_j \right\|^2 \\
&= \left\langle \sum_{j=1}^{p} y_j \otimes v_j \,,\, \sum_{k=1}^{p} y_k \otimes v_k \right\rangle \\
&= \sum_{j,k} \langle y_j \otimes v_j \,,\, y_k \otimes v_k \rangle \\
&= \sum_{j,k} \langle y_j \,,\, y_k \rangle \, \langle v_j \,,\, v_k \rangle \\
&= \sum_{j} \langle y_j \,,\, y_j \rangle \\
&= \sum_{j} \|y_j\|^2
\end{aligned}
$$

(another way to see this is to observe that Y is deduced from A by a rotation, which is an isometry, and then $\|A\| = \|Y\|$). As $Y = U\Sigma$ with $U^{\mathsf{T}}U = \mathbb{I}_p$ and the diagonal terms of Σ being $(\sigma_j)_j$, we have

$$\|y_j\| = \sigma_j. \tag{2.6.6}$$

Hence, the norm of A can be partitioned as

$$\|A\|^2 = \sum_j \sigma_j^2. \tag{2.6.7}$$

Let us recall that

$$\lambda_j = \sigma_j^2, \qquad \text{with} \quad A^{\mathsf{T}}Av_j = \lambda_j v_j.$$

Then

$$\|A\|^2 = \sum_{i=1}^{p} \lambda_i, \tag{2.6.8}$$

and the quality of the representation of A by its projection on the axis spanned by v_j is

$$\varrho_j = \frac{\|y_j\|^2}{\|A\|^2} = \frac{\lambda_j}{\sum_i \lambda_i}. \tag{2.6.9}$$

The quality of representation of the point cloud (i.e., of array A) by its projection A_r on the subspace spanned by vectors $(v_1, \ldots, v_r)$ is

$$\begin{aligned} \rho_r &= \sum_{j=1}^{r} \varrho_j \\ &= \frac{\sum_{j=1}^{r} \lambda_j}{\sum_i \lambda_i}. \end{aligned} \tag{2.6.10}$$

The quality of representation of item i on axis $j \in \{1, p\}$ is

$$\psi(i, j) = \frac{y_{ij}^2}{\sum_{\ell=1}^{p} y_{i\ell}^2}, \tag{2.6.11}$$

and the quality of projection of item i on the subspace spanned by vectors $(v_1, \ldots, v_r)$ is

$$\begin{aligned} \theta(i, r) &= \sum_{j=1}^{r} \psi(i, j) \\ &= \frac{\sum_{j=1}^{r} y_{ij}^2}{\sum_{\ell=1}^{p} y_{i\ell}^2}. \end{aligned} \tag{2.6.12}$$

This can be summarized as

Quality of representation of	Notation	Calculation
Item i on axis j	$\psi(i, j)$	$\dfrac{y_{ij}^2}{\sum_{\ell=1}^{p} y_{i\ell}^2}$
Item i on subspace E_r	$\theta(i, r)$	$\sum_{j=1}^{r} \psi(i, j)$
Point cloud on axis j	ϱ_j	$\dfrac{\lambda_j}{\sum_{i=1}^{p} \lambda_i}$
Point cloud on subspace E_r	ρ_r	$\sum_{j=1}^{r} \varrho_j$

- **Prescribed rank or accuracy:** In the algebraic framework, PCA is about the best low rank approximation of a matrix. This can be set in two guises:

  $\rightarrow$ Select a rank r, and deduce the quality $\rho(r) = \rho_r$ of the approximation.
  $\rightarrow$ Select a quality ρ of an approximation, and deduce the smallest rank r at which it should be done, by

$$r = \min_{1 \le r < p} \{r \mid \rho_r \ge \rho\}.$$

The former is PCA at *prescribed rank*, whereas the latter is PCA at *prescribed accuracy*. The key tool for implementing the one or the other is the curve $r \mapsto \rho(r)$.

2.7 Classical Analysis

Here, we denote $A \ge 0$ if all coefficients in A are nonnegative. Let us recall that $a_i \in \mathbb{R}^p$; a_i denotes the row i of $A \in \mathbb{R}^{n \times p}$ and $a_{*j} \in \mathbb{R}^n$ its column j.

The mean and standard deviation of a distribution are often the best summary of it. If PCA yields the best rank one approximation, it is likely that first axis and components mirror this best summary and bring few information on the inner structure of matrix A. Hence a standard procedure is to center and scale a dataset and run the PCA on the scaled and centered dataset to focus on the inner structure (e.g., correlations between columns).

- **Centering:** Centering a matrix A is translating the attached point cloud $\mathcal{A}$ to its barycenter:

$$\text{in } \mathbb{R}^p, \qquad a_i \xrightarrow{\text{centering}} \overline{a_i} = a_i - g, \tag{2.7.1}$$

where $g \in \mathbb{R}^p$ is the barycenter of the point cloud, i.e.,

$$g_j = \frac{1}{n} \sum_i a_{ij}. \tag{2.7.2}$$

It is easy to check that $\sum_i \overline{a_i} = \sum_i \left(a_i - \frac{1}{n} \sum_i a_i \right) = \sum_i a_i - \sum_i a_i = 0$. Matrix $\overline{A}$ is centered column-wise:

$$\forall j, \quad \sum_i \overline{a}_{ij} = 0. \tag{2.7.3}$$

- **Scaling:** Scaling a matrix column-wise is dividing each column vector a_{*j} by its norm, aka its standard deviation if it is centered:

$$a_{*j} \xrightarrow{\text{scaling}} \frac{a_{*j}}{\|a_{*j}\|}. \tag{2.7.4}$$

The centered-scaled matrix A' is defined by

$$a_{*j} \longrightarrow \frac{\overline{a}_{*j}}{\|\overline{a}_{*j}\|} \quad \text{with} \quad \overline{a}_{*j} = a_{*j} - g_j \mathbf{1}_n. \tag{2.7.5}$$

Scaled-centered PCA of a matrix A is defined as follows:

Algorithm 5 PCA-SC(A)

1: **input** $A \in \mathbb{R}^{n \times p}$
2: **compute** the barycenter of A: $g = \frac{1}{n} \sum_i a_i$
3: **center** $A : A \longrightarrow \overline{A}$, with $\forall\, i, \quad a_i \longrightarrow \overline{a}_i = a_i - g$
4: **scale** $\overline{A}: \overline{A} \longrightarrow A'$, with $\forall\, j, \quad \overline{a}_{*j} \longrightarrow a'_{*j} = \frac{\overline{a}_{*j}}{\|\overline{a}_{*j}\|}$
5: **do** $Y, \Lambda, V = $ PCA_CORE(A')
6: **return** Y, Λ, V

There are a few elementary results for scaled-centered PCA. By definition, the coefficients $c_{j\ell}$ of $C = A'^{\mathsf{T}} A'$ are the correlations between centered-scaled variables a'_{*j} and $a'_{*\ell}$. Hence, we have

$$-1 \le c_{j\ell} \le 1. \tag{2.7.6}$$

We also have

$$\forall\, j, \qquad c_{jj} = 1. \tag{2.7.7}$$

Hence

$$\sum_j \lambda_j = \operatorname{Tr} C = p. \tag{2.7.8}$$

Hence, the quality of approximation at rank r of A' is

$$\rho_r = \frac{1}{p} \sum_{j \le r} \lambda_j. \tag{2.7.9}$$

- **Double averaging or bi-centering:** Let us have a matrix $A \ge 0$ which is, for example, an array of counts. A classic example is a contingency table (contingency tables can be analyzed with Correspondence Analysis, see Chap. 5). The structure of A is dominated by the property $A \geq 0$. If a PCA of A is run, this will be the main (trivial) information given by axis 1. This trivial information can be filtered out by setting the model

$$a_{ij} = \underbrace{m}_{\text{global mean}} + \underbrace{x_i}_{\text{effect of row } i} + \underbrace{y_j}_{\text{effect of column } j} + \underbrace{r_{ij}}_{\text{residuals}} , \tag{2.7.10}$$

with

$$\begin{cases} \sum_i x_i = 0 \\ \sum_j y_j = 0 \\ \forall j, \quad \sum_i r_{ij} = 0 \\ \forall i, \quad \sum_j r_{ij} = 0. \end{cases} \tag{2.7.11}$$

Then, we have

$$\begin{cases} m = \dfrac{1}{np} \sum_{i,j} a_{ij} \\ x_i = \left(\dfrac{1}{p} \sum_j a_{ij} \right) - m \\ y_j = \left(\dfrac{1}{n} \sum_i a_{ij} \right) - m. \end{cases} \tag{2.7.12}$$

Indeed, we have

$$\sum_{i,j} a_{ij} = np\, m,$$

so

$$m = \frac{1}{np} \sum_{ij} a_{ij}.$$

Then

$$\sum_i a_{ij} = n\, m + n y_j,$$

and

$$y_j = \left(\frac{1}{n} \sum_i a_{ij} \right) - m. \tag{2.7.13}$$

Similarly

$$\sum_j a_{ij} = p\,m + p\,x_i,$$

and

$$x_i = \left(\frac{1}{p} \sum_j a_{ij} \right) - m. \tag{2.7.14}$$

So, we have

$$
\begin{aligned}
a_{ij} &= m + x_i + y_i + r_{ij} \\
&= m + \left(\frac{1}{p} \sum_j a_{ij} \right) - m + \left(\frac{1}{n} \sum_i a_{ij} \right) - m + r_{ij} \\
&= -\frac{1}{np} \sum_{i,j} a_{ij} + \left(\frac{1}{p} \sum_j a_{ij} \right) + \left(\frac{1}{n} \sum_i a_{ij} \right) + r_{ij},
\end{aligned} \tag{2.7.15}
$$

which is denoted as well

$$r_{ij} = a_{ij} - a_{i.} - a_{.j} + a_{..} \tag{2.7.16}$$

PCA with double averaging is:

1. Compute the global mean m, each effect x_i and y_j, and the matrix of residuals R.
2. Run the PCA of R, which is already centered, without scaling

Chapter 3
Complements on PCA

Abstract In this chapter, we present some additional information on PCA. After a reminder of some basic statistical concepts, the statistical approach to PCA is developed. This clarifies an often obscure point, a source of much discussion, on the links between PCA and Factor Analysis, in particular via the notion of probabilistic PCA. It is possible to propose a statistical approach for each of the methods in this book, but, with the exception of Correspondence Analysis, this aspect will not be developed beyond PCA. We then present a very rich topic that is often omitted in texts on PCA: the characterization of the eigenvalue distributions of random matrices, in particular the Wishart distribution and the Marčenko-Pastur semicircular law. PCA relies on the Euclidean distance to compare matrices or point clouds. An extension to other distances is still a topical but difficult issue. In vector spaces, distances are related to a norm. Such an extension has been developed in the case of Unitarily Invariant Norms, which we discuss here.

3.1 Preliminaries

A sample is a collection of outcomes of a random experiment, like rolling a dice. In this example, the set of possible outcomes is $\mathcal{X} = \{1, 2, 3, 4, 5, 6\}$, and a sample with $n = 5$ repetitions is, e.g., $S = (1, 5, 3, 1, 4)$. A random variable, usually denoted by capital letters, like X, is a probability distribution on the outcomes of a random experiment, i.e., on $\mathcal{X}$. It is denoted

$$p(x) = \mathbb{P}(X = x), \qquad x \in \mathcal{X}.$$

In dice rolling, the random variable X is defined by

$$p_1 = \mathbb{P}(X = 1), \quad p_2 = \mathbb{P}(X = 2), \quad \ldots \tag{3.1.1}$$

Let the dice roll n times. The sample is a tuple in $\{1, \ldots, 6\}^n$, referred to as n realizations of X. They are called i.i.d. or independent identically distributed. The

A. Franc, *Linear Dimensionality Reduction*, Lecture Notes in Statistics 228, https://doi.org/10.1007/978-3-031-95785-7_3

notion of population is more subtle. Let us consider that it is an enumerable sample, i.e., $n = +\infty$.

Let X be a random variable with values in $\mathbb{R}$. Let $x = (x_i)_i$ with $1 \leq i \leq n$ be a vector of n realizations of X (a sample of size n). Then, the sample mean is

$$\overline{x} = \frac{1}{n} \sum_{i=1}^{n} x_i. \tag{3.1.2}$$

It must not be confounded with the population mean, which is the expectation of the r.v. X, denoted $\mathbb{E}(X)$. We then have

$$\left| \begin{array}{ll} \mathbb{E}(X) & \text{the population mean} \\ \overline{x} & \text{the sample mean.} \end{array} \right.$$

A sample is a subset of a possibly infinite population. Statistics are about inferring properties for the population knowing a sample. For example, if an infinite population follows a Gaussian law of mean μ and variance σ^2, the population mean is μ, whereas the sample mean is $\overline{x} = (1/n) \sum_i x_i$. The sample mean is an unbiased estimator of the population mean.

The population variance, or the variance of the r.v., is defined as

$$\text{var } X = \mathbb{E}\left((X - \mathbb{E}(X))^2 \right). \tag{3.1.3}$$

The sample variance is defined as

$$\text{var } x = \frac{1}{n-1} \sum_{i=1}^{n} (x_i - \overline{x})^2. \tag{3.1.4}$$

It is an unbiased estimator of the population variance. It is easy to check that

$$\text{var } X = \mathbb{E}(X^2) - (\mathbb{E}(X))^2. \tag{3.1.5}$$

Then, if X is centered ($\mathbb{E}(X) = 0$), we have

$$\text{var } X = \mathbb{E}(X^2). \tag{3.1.6}$$

The standard deviation is the square root of the variance

$$\text{std } x = \sqrt{\text{var } x} = \sqrt{\frac{1}{n-1} \sum_{i=1}^{n} (x_i - \overline{x})^2}. \tag{3.1.7}$$

The covariance of two random variables X and Y is

$$\text{cov}\,(X, Y) = \mathbb{E}((X - \mathbb{E}(X))(Y - \mathbb{E}(Y))), \qquad (3.1.8)$$

and

$$\text{cov}\,(x, y) = \frac{1}{n-1} \sum_{i=1}^{n} (x_i - \overline{x})(y_i - \overline{y}). \qquad (3.1.9)$$

Considering that $x = (x_1, \ldots, x_n) \in \mathbb{R}^n$ and similarly for y, it is a symmetric, semi-definite, positive bilinear form on $\mathbb{R}^n$. It is definite, hence an inner product, for those samples with $\overline{x} = \overline{y} = 0$: The map $(x, y) \longrightarrow \text{cov}\,(x, y)$ is an inner product, and $x \longrightarrow \text{std}\,x$ is a norm, proportional to the Frobenius norm. Independence is orthogonality: Two random variables are independent if $\text{cov}\,(X, Y) = 0$. In such a case

$$\text{var}\,(X + Y) = \text{var}\,X + \text{var}\,Y. \qquad (3.1.10)$$

This is Pythagoras theorem. Indeed, more generally,

$$\begin{aligned}
\text{var}\,(X + Y) &= \mathbb{E}\left\{[X + Y - \mathbb{E}(X + Y)]^2\right\} \\
&= \mathbb{E}\left\{[(X - \mathbb{E}(X)) + (Y - \mathbb{E}(Y))]^2\right\} \\
&= \mathbb{E}\left\{[X - \mathbb{E}(X)]^2 + [Y - \mathbb{E}(Y)]^2 + 2\,[(X - \mathbb{E}(X))(Y - \mathbb{E}(Y))]\right\} \\
&= \text{var}\,X + \text{var}\,Y + 2\,\text{cov}\,(X, Y).
\end{aligned}$$

$$(3.1.11)$$

For the sake of simplicity for introducing those notions, we have selected a discrete random variable, like the distribution of the outcomes of dice rolling. An r.v. can be discrete and countable, like a Poisson distribution of parameter λ:

$$\mathbb{P}(X = k) = \frac{\lambda^k}{k!}\,e^{-\lambda},$$

or continuous, like a Gaussian law

$$p(x) = \frac{1}{\sigma\sqrt{2\pi}}\,\exp -\frac{1}{2}\left(\frac{x - \mu}{\sigma}\right)^2.$$

($\mathbb{P}(X = x)$ is meaningless for a continuous r.v., where p is the probability density function, i.e., $\mathbb{P}(X \in [a, b]) = \int_a^b p(x)\,dx$.) It can be multivariate too (i.e., with values in $\mathbb{R}^p$), like the multivariate Gaussian distribution $\mathcal{N}(\boldsymbol{\mu}, \Sigma)$:

$$p(\mathbf{x}) = \frac{1}{(2\pi)^{p/2}}\frac{1}{\sqrt{\det \Sigma}}\,\exp -\frac{1}{2}(\mathbf{x} - \boldsymbol{\mu})^{\mathsf{T}}\Sigma^{-1}(\mathbf{x} - \boldsymbol{\mu}),$$

with $\mathbf{x} = (x_1, \ldots, x_n)$, $\boldsymbol{\mu} = (\mu_1, \ldots, \mu_n)$, and Σ is the $p \times p$ variance-covariance matrix, i.e., $\Sigma_{ij} = \mathrm{cov}\,(X_i, X_j)$.

Here is a simple observation that will be useful in next sections. Let us have a p-variate Gaussian r.v. with pdf given by $\mathcal{N}(0, \Sigma)$, i.e., $\boldsymbol{\mu} = 0$, and n realizations of it. Each realization is a vector in $\mathbb{R}^p$. The sample of the n realizations can be described as a matrix $A \in \mathbb{R}^{n \times p}$, with row a_i being the i-th realization. Columns a_{*j} are realizations of random variables X_j, respectively. Let us assume that matrix A is column-wise centered. Then, an (biased) estimate of $\mathrm{cov}(X_j, X_k)$ is $\mathrm{cov}(X_j, X_k) = \frac{1}{n}\sum_{i=1}^{n} a_{ij}a_{ik}$, or, in a more compact way, an estimate of Σ is

$$\widetilde{\Sigma} = \frac{1}{n}\,A^{\mathsf{T}}A. \tag{3.1.12}$$

An unbiased estimator would be $\frac{1}{n-1}A^{\mathsf{T}}A$, but as calculations are often made when n is large, and even, for asymptotic regime, $n \to \infty$, it is often standard to keep $(1/n)A^{\mathsf{T}}A$.

3.2   Statistical Approach

The statistical approach to PCA is a complex topic with many guises. Some rely on empirical distributions, without inference, and some rely on inference, with or without latent variables. Some are presented here.

Statistical approach has been set first with datasets being n i.i.d. realizations of a random variable. Let us have a p-variate random variable (r.v.) $X = (X_1, \ldots, X_p)$ with zero mean: $\mathbb{E}(X) = (0, \ldots, 0)$, observed on n independent items. The reason for considering r.v. with zero mean is that we are interested in decomposing the structure of covariance between the X_j. A typical example is a p-variate Gaussian distribution with zero mean and Σ as variance-covariance matrix. The n observations for a variable X_j are the column j of a matrix $A \in \mathbb{R}^{n \times p}$. Each variable X_j is centered: $\mathbb{E}(X_j) = 0$. PCA at rank r is about finding r independent (non-correlated) new r.v. $(Y_k)_k$, each a linear combination of the X_j, $Y_k = \sum_{j=1}^{p} v_{kj}X_j$, with $v_k \in \mathbb{R}^p$, $1 \le k \le r$, which have maximum variance under the constrain $\|v_k\| = 1$. Here, means, variances, and covariances are sample mean, sample variances, and sample covariances.

Let us consider A as a sample of n realizations of X, with sample variance-covariance $C = \frac{1}{n}A^{\mathsf{T}}A$. Because $\mathbb{E}(X) = 0$, we assume that A is column-wise centered, i.e., $\sum_i a_{ij} = 0$ for all j. Let $v \in \mathbb{R}^p$, and let us consider the variable

$Y = \sum_j v_j X_j$, or $y_i = \sum_j v_j a_{ij}$. It is centered as each a_{*j} is centered (or $\mathbb{E}(Y) = \sum_j v_j \mathbb{E}(X_j) = 0$), and its variance is

$$\text{var } Y = \frac{1}{n} \sum_i y_i^2, \qquad \text{as } Y \text{ is centered}$$

$$= \frac{1}{n} \langle Av , \, Av \rangle$$

$$= \frac{1}{n} \langle v, A^{\mathrm{T}} Av \rangle$$

$$= \langle v, Cv \rangle.$$

Maximizing var Av with the constraint $\|v\| = 1$ yields that v is the eigenvector of C associated with its largest eigenvalue, denoted λ. Then, as $Cv = \lambda v$, var $Av = \langle v, Cv \rangle = \lambda \langle v, v \rangle = \lambda$.

The general result is as follows: Let X be a random variable on $\mathbb{R}^p$, with $\mathbb{E}(X) = 0$. Let us denote var $X = \Sigma$. Then, there exists a rotation $U \in \mathbb{O}(\mathbb{R}^p)$ defining

$$V = XU$$

such that the covariance matrix of V is diagonal and the k-th component of V has maximum variance among all normalized linear combinations uncorrelated with $V_1, \ldots, V_{k-1}$.

Notes and References A standard textbook for statistical approach to PCA is [And58]. This section is adapted from [And58, chap. 11]. The general result is [And58, Th. 11.2.1]. Another key reference is [Rao64].

3.3 Factor Analysis and Probabilistic PCA

Let us have a p-variate Gaussian variable, with n i.i.d. realizations, considered as observations. Correlations between observed variables are expressed by the variance-covariance matrix Σ. Factor analysis (FA) models a situation where there exist r independent unobserved variables, with $r < p$, the response to which explains the correlation between the observed variables. It is a latent variable statistical model. Unobserved variables are called *hidden variables* or *latent variables*. Let z denote the latent variable and x the observed variable (the data). Let us assume that random variable Z follows a certain law, $\mathbb{P}(Z = z \mid \theta)$, denoted

$$p(z \mid \theta),$$

which will be specified later. Then, the model for X in FA can be written

$$p(x \mid z, \theta'),$$

and solving FA is about inferring parameters in θ, θ' and recovering the latent variables Z.

This makes FA rather complex, and this complexity has nourished decades of debates about Factor Analysis and PCA. Therefore, many authors (including [Jol02, chap. 7]) have insisted on the difference between PCA and FA. In fact, FA is a statistical approach of PCA where principal axes (and their realizations as components) are latent variables.

Let us select for the latent variables a multivariate Gaussian probability

$$p(z \mid \theta) = \mathcal{N}(z \mid \mu_0, \Sigma_0), \tag{3.3.1}$$

(here, $\theta = (\mu_0, \Sigma_0)$). Then, we select for $p(x \mid z)$ a Gaussian law as well, denoted

$$p(x \mid z, \theta') = \mathcal{N}(Wz + \mu, \Psi) \tag{3.3.2}$$

($\theta' = (W, \mu, \Psi)$), where the mean has been selected as a linear function of the hidden input ($x_i \in \mathbb{R}^p$, $z \in \mathbb{R}^r$, and $W \in \mathbb{R}^{p \times r}$) and Ψ to be diagonal. The special case where $\Psi = \sigma^2 \mathbb{I}_p$ is known as *probabilistic PCA*. If $\sigma \to 0$, probabilistic PCA converges to PCA which is non-probabilistic. We then have the following reductions:

$$\text{FA: } \Psi \xrightarrow{\Psi = \sigma^2 \mathbb{I}} \text{ probabilistic PCA } \xrightarrow{\sigma = 0} \text{ PCA.}$$

Marginalizing over the latent variable leads to

$$\begin{aligned} p(x \mid \theta) &= \int_z p(x \mid z)\, p(z \mid \theta)\, dz \\ &= \mathcal{N}(x \mid W\mu_0 + \mu, \, \Psi + W\Sigma_0 W^{\mathsf{T}}). \end{aligned} \tag{3.3.3}$$

This shows that it is possible to choose $\mu_0 = 0$ and $\Sigma_0 = \mathbb{I}_r$ without loss of generality:

$$p(z \mid \theta) = \mathcal{N}(0, \mathbb{I}_r), \tag{3.3.4}$$

which leads to

$$p(x \mid z, \theta') = \mathcal{N}(Wz + \mu, \, WW^{\mathsf{T}} + \Psi), \tag{3.3.5}$$

and

$$p(x \mid \theta) = \mathcal{N}(x \mid \mu, \, WW^{\mathsf{T}} + \Psi). \tag{3.3.6}$$

The link with PCA can be seen in the observation that the variance-covariance matrix of the data x_i is given by a low-rank matrix $WW^{\mathsf{T}} + \Psi$, as $W \in \mathbb{R}^{p \times r}$ and rank $WW^{\mathsf{T}} = r$. The variance-covariance structure of the observed variables is split into a variance structure for each variable given by Ψ and the covariance structure given by W.

The next step is to estimate the parameters (W, μ, σ) or (W, μ, Ψ) of the model knowing the observations, not presented here. This is not so easy and is presented in [Bis06, sect. 12.2.1]. It can be done with the EM procedure (see Mur12, sec. 12.1.5).

Notes and References Factor analysis has been developed all over the twentieth century in parallel with PCA, sometimes with some controversies. One probable reason for those controversies is that sometimes it is said to inherit from the work of Pearson in 1901, and sometimes of Spearman in 1904, who was interested in finding one factor explaining a diversity of correlated features. His work has been extended to multivariate factors by Thurstone in 1935. The coexistence of PCA and FA with the same tools has probably contributed to some controversies. A history of FA can be found in the introduction of [Bas94]. A sound use of notions in statistical modeling has clarified the situation. A classical presentation with this approach is [And58, sect. 4.7] or [Bas94]. It is currently developed as a statistical model with latent variables, where the correlated features are the observed variables and the independent factors the latent variables. Classical setting is with Gaussian models. Probabilistic PCA (PPCA) has been proposed in 1999 in [TB99], which has presented also the links and differences between PCA, PPCA, and FA. See also [Bis06, sect. 12.2] for a clear presentation of statistical approach of PCA, probabilistic PCA, and factor analysis. We have followed [Bis06] and [Mur12, chap. 12] here.

3.4 Distribution of Eigenvalues of Random Matrices

A random matrix is the realization of a random variable over a space of matrices. Different spaces have been studied and classically referred to for historical reasons as "ensembles." For example, the Ginibre ensemble is the set of $n \times n$ complex matrices whose entries are i.i.d. realizations of the complex Gaussian normal law, i.e., $a_{k\ell} = x_{k\ell} + i\, y_{k\ell}$ with $x_{k\ell}, y_{k\ell} \sim \mathcal{N}(0, 1)$. There are several types of results about the distribution of the eigenvalues of random matrices, like:

→ What is the distribution of the eigenvalues of a matrix in a given ensemble for size n?

→ When $n \to \infty$, does the distribution of the eigenvalues converge to a given "universal" distribution ?

Both issues are addressed in this section for variance-covariance matrices of Gaussian random matrices, with:

◇ The distribution of eigenvalues of a Wishart matrix
◇ Their asymptotic distribution when $n \to \infty$

Wishart Matrix Let $A \in \mathbb{R}^{n \times p}$ be a random matrix, with rows being i.i.d. realizations of a p-variate Gaussian distribution X of mean $\mu = 0$ and variance-covariance matrix Σ. Let us denote $C = A^{\mathrm{T}} A \in \mathbb{R}^{p \times p}$. Then, elements in C follow a Wishart distribution, denoted $W(n, p, \Sigma)$, which has been analytically computed by Wishart in 1928. For $\Sigma = \mathbb{I}_p$, it is given by

$$p(C) = w(n, p) \, (\det C)^{(n-p-1)/2} \, \exp -\frac{1}{2}\mathrm{Tr}\, C, \qquad (3.4.1)$$

where $w(n, p)$ is the normalizing constant

$$\frac{1}{w(n, p)} = \pi^{p(p-1)/4}\, 2^{np/2} \prod_{j=1}^{p} \Gamma\left(\frac{n - j + 1}{2}\right). \qquad (3.4.2)$$

The formula for a general Σ is

$$p(C) = w(n, p, \Sigma) \, (\det C)^{(n-p-1)/2} \, \exp -\frac{1}{2}\mathrm{Tr}\, \Sigma^{-1} C, \qquad (3.4.3)$$

with

$$\frac{1}{w(n, p, \Sigma)} = \pi^{p(p-1)/4}\, 2^{np/2}\, (\det \Sigma)^{n/2} \prod_{j=1}^{p} \Gamma\left(\frac{n - j + 1}{2}\right). \qquad (3.4.4)$$

Marčenko-Pastur Semicircular Law This laws gives the asymptotic distribution of eigenvalues of a Wishart matrix for large matrices in the form of their probability density function ($\mathbb{P}(\lambda \in [x, y])$ with $[x, y] \subset \mathbb{R}$). Let $A \in \mathbb{R}^{n \times p}$ be a random matrix with its elements being i.i.d. realizations of a normal distribution: $a_{ij} \sim \mathcal{N}(0, 1)$. Let

$$C = \frac{1}{n} A^{\mathrm{T}} A$$

be the variance-covariance matrix of A. Let us denote by

$$\lambda_1 > \ldots \lambda_p > 0$$

the p eigenvalues of C, assumed to be separated. Marčenko and Pastur consider the case where $n \to +\infty$ with p controlled by n, i.e., n and p grow with

$$n \to \infty, \qquad \lim_{n\to\infty} \frac{p(n)}{n} = \alpha.$$

Let us assume that $0 < \alpha \leq 1$, and define

$$\begin{cases} a & = (1 - \sqrt{\alpha})^2 \\ b & = (1 + \sqrt{\alpha})^2. \end{cases}$$

A first result is that a and b are a good approximation of the smallest and the largest eigenvalue of C for large n, respectively. The range of eigenvalues of C increases with α, as $b - a = 4\sqrt{\alpha}$. Let μ be the measure such that

$$\mu(\Omega) = \frac{1}{p}\#\{i \mid \lambda_i \in \Omega\} \tag{3.4.5}$$

for Ω being, to keep it simple, an interval in $\mathbb{R}$ (the eigenvalues of C are real). Then, asymptotically for $n \to \infty$,

$$\lambda \in [a, b], \tag{3.4.6}$$

and, if $x \in [a, b]$,

$$d\mu(x) = \frac{1}{2\pi\alpha} \frac{\sqrt{(b - x)(x - a)}}{x}. \tag{3.4.7}$$

So, if $a \leq z < z' \leq b$,

$$\mathbb{P}(z \leq \lambda \leq z') = \frac{1}{2\pi\alpha} \int_z^{z'} \frac{\sqrt{(b - x)(x - a)}}{x}\, dx. \tag{3.4.8}$$

Let us note that the cumulative distribution function of the eigenvalues, denoted $f(y) = \mathbb{P}(\lambda \leq y)$, is an elliptic function:

$$f(y) = \frac{1}{2\pi\alpha} \int_a^y \frac{\sqrt{(b - x)(x - a)}}{x}\, dx. \tag{3.4.9}$$

Notes and References   The Wishart matrix has been thoroughly studied, and much can be said about it. See, e.g., [And58, chapter 7] devoted to it. The expressions of the joint law of the coefficients of a Wishart matrix in the case where $\Sigma = \mathbb{I}_p$ and for any Σ have been borrowed from [And58, section 7.2, formula (1)] and [Mec19, section 2.2]. For the asymptotic distribution of the eigenvalues of a Wishart matrix through the Marčenko-Pastur theorem, we have followed [Mec20], which is a gem.

3.5 Unitarily Invariant Norms

The problem of PCA as set in Sect. 2.1 can be set for any norm and not Frobenius norm only. Most widely used norms in data analysis are ℓ^1 and ℓ^∞ norms, on top of ℓ^2 norms. The spectral norm of a matrix $A \in \mathbb{R}^{n \times p}$ is

$$\|A\|_{\mathrm{sp}} = \max_{\|x\|=1} \|A^T A x\|.$$

However, there are very few norms for which exact solution and efficient algorithms to compute a solution are known. One exception is the family of unitarily invariant norms, for which there is a generalization of Eckart-Young theorem. The framework is that of vector spaces on $\mathbb{C}$, but it can be applied on vector spaces on $\mathbb{R}$ as well. $\mathbb{U}(\mathbb{C}^n)$ or $\mathbb{U}(n)$ denotes the set of unitary matrices in $\mathbb{C}^n$, i.e., matrices having the property $UU^* = U^*U = \mathbb{I}_n$, where $U^* = \overline{U^T}$ (and the same for $\mathbb{C}^p$). The equivalent in $\mathbb{R}^n$ is the set $\mathbb{O}(\mathbb{R}^n)$ of orthogonal matrices such that $U^T U = \mathbb{I}_n$.

Let E, F be two complex vector spaces, $A \in E \otimes F \simeq \mathcal{L}(F, E)$. A norm $\|.\|$

$$E \otimes F \xrightarrow{\ \|\cdot\|\ } \mathbb{R}^+$$

is said unitarily invariant if

$$\forall \begin{cases} U & \in \mathbb{U}(\mathbb{C}^n) \\ V & \in \mathbb{U}(\mathbb{C}^p)\,, \\ A & \in \mathbb{C}^{n \times p} \end{cases} \qquad \|U A V^*\| = \|A\|. \tag{3.5.1}$$

For example, spectral and Frobenius norms are unitarily invariant norms. If E, F are real vector spaces, the norm is said invariant by orthogonal transformation if

$$\forall \begin{cases} U & \in \mathbb{O}(\mathbb{R}^n) \\ V & \in \mathbb{O}(\mathbb{R}^p)\,, \\ A & \in \mathbb{R}^{n \times p} \end{cases} \qquad \|U A V^T\| = \|A\|. \tag{3.5.2}$$

A norm

$$\mathbb{R}^n \xrightarrow{\ \Phi\ } \mathbb{R}^+$$

is called a Symmetric Gauge Function (SGF) if it is invariant by any permutation of the coordinates in $\mathbb{R}^n$, i.e., if $\mathscr{P}$ is the set of permutations in $\mathbb{R}^n$

$$\forall P \in \mathscr{P}, \quad \Phi(Px) = \Phi(x). \tag{3.5.3}$$

There is a remarkable link between UIN and SGF. Let $A \in \mathbb{C}^{n \times p}$ (or $\in \mathbb{R}^{n \times p}$) and

$$\Sigma = (\sigma_1, \ldots, \sigma_p)$$

be the set of its singular values. To each SGF Φ, one associates the norm $\|.\|_\Phi$ defined by

$$\|A\|_\Phi = \Phi(\sigma_1, \ldots, \sigma_n). \tag{3.5.4}$$

Then, $\|.\|_\Phi$ is a UIN. Let us note that if $\|.\|$ is a UIN, A is a matrix in $\mathbb{C}^{n \times p}$, and $A = U\Sigma V^*$ is the SVD of A, then $\|A\| = \|U^*AV\|$ and, as $U^*AV = \Sigma$, $\|A\| = \|\Sigma\|$, i.e., is a function of its singular values.

- **Mirsky's theorem:** Let E, F be two vector spaces on $\mathbb{C}$ or $\mathbb{R}$, and $A, B \in E \otimes F \simeq \mathcal{L}(F, E)$. Let $(\alpha_1, \ldots, \alpha_p)$ (resp. $(\beta_1, \ldots, \beta_p)$) be the singular values of A (resp. B). Mirsky has shown that, for any unitarily invariant norm $\|.\|$,

$$\|\text{diag}\,(\alpha_1 - \beta_1, \ldots, \alpha_p - \beta_p)\| \leq \|A - B\|. \tag{3.5.5}$$

- **Schmidt-Mirsky theorem:** Let:

  $\rightarrow \quad A \in E \otimes F \simeq \mathcal{L}(F, E)$.
  $\rightarrow \quad (\sigma_1, \ldots, \sigma_p)$ be the singular values of A in nonincreasing order.
  $\rightarrow \quad B \in E \otimes F$ with rank $B = r \leq p$.
  $\rightarrow \quad \Phi$ be a Symmetric Gauge Function with $\|.\|_\Phi$ as associated unitarily invariant norm.

Then

$$\|A - B\|_\Phi \geq \Phi(0, \ldots, 0, \sigma_{r+1}, \ldots, \alpha_p). \tag{3.5.6}$$

Moreover, if $A = U\Sigma V^*$ is the SVD of A, the equality is reached for matrix A_r defined as

$$A_r = U\Sigma_r V^*, \tag{3.5.7}$$

where Σ_r is the diagonal matrix obtained from Σ be setting to 0 all singular values beyond r.

Then, A_r is the best rank r approximation of A for norm $\|.\|_\Phi$ as well.

Notes and References The link between UIN and SGF has been shown in J. von Neumann (1937), Some Matrix Inequalities and Metrization of Matrix Spaces, *Tomsk Univ. Rev.*, **1286–300**. The extension of PCA to UIN and SGF is nicely presented with many references in [Cla87]. See [Sch60] as well for an algebraic survey.

Chapter 4
PCA with Metrics on Rows and Columns

Abstract In this chapter, we show how PCA can be developed in a space endowed with a Euclidean structure. The space studied is the vector space of matrices of a given dimension, on which PCA is run as a best approximation with a matrix of the same dimensions but lower rank. The term "best approximation" is to be understood as the approximation by the smallest distance once a Euclidean structure has been chosen via an inner product. We focus on Euclidean structures built in a consistent way with Euclidean structures in the vector space spanned by the columns and by the rows, which specify how to compare variables and items. We then develop the geometric approach to this problem, where the best projections are derived with metrics chosen on these spaces. This chapter is crucial because several of the methods presented in later chapters can be seen as PCA with specific Euclidean metrics. Thus, their solution can be built simply by translating the solution provided in this chapter.

A matrix $A \in \mathbb{R}^{n \times p}$ being given, the algebraic setting of PCA at rank r, is about finding the matrix $A_r \in \mathbb{R}^{n \times p}$ of rank r which is the closest to A, i.e., such that $\|A - A_r\|$ is minimal. Implicitly, in this definition given in Chap. 2, the distance is the one induced by the Frobenius norm, itself induced by the inner product. If $A = (a_{ij})_{i,j}$ and $B = (b_{ij})_{i,j}$, we have

$$d(A, B) = \|A - B\| = \sqrt{\sum_{i,j} (a_{ij} - b_{ij})^2}.$$

This distance is associated with the inner product $\langle ., . \rangle$ between matrices ($\mathbb{R}^{n \times p}$ is a vector space) defined by

$$\forall\, A, B \in \mathbb{R}^{n \times p}, \quad \langle A, B \rangle = \sum_{i,j} a_{ij} b_{ij}$$

$$= \mathrm{Tr}\, A^{\mathrm{T}} B$$

$$= \mathrm{Tr}\, B^{\mathrm{T}} A,$$

A. Franc, *Linear Dimensionality Reduction*, Lecture Notes in Statistics 228,
https://doi.org/10.1007/978-3-031-95785-7_4

where Tr is the trace of a matrix

$$\operatorname{Tr} A = \sum_i a_{ii}.$$

But it is possible to define and select other distances in $\mathbb{R}^{n \times p}$. We focus in this chapter on distances induced by Euclidean structures in the space of matrices, because the extension of PCA to these situations is straightforward. The notion and definition of a Euclidean structure in a real vector space are recalled in Appendix B.2, especially Sect. B.2.

4.1 Euclidean Structure on the Space of Matrices $\mathbb{R}^{n \times p}$

Let us assume that an inner product is given in space $E = \mathbb{R}^n$ by a Symmetric Definite Positive (SDP) matrix $N \in \mathbb{R}^{n \times n}$ and in space $F = \mathbb{R}^p$ by an SDP matrix $P \in \mathbb{R}^{p \times p}$. Here, we build an inner product in $\mathbb{R}^{n \times p}$ consistent with them. We define $M \in \mathbb{R}^{n \times n}$ (respectively, $Q \in \mathbb{R}^{p \times p}$) as the unique SDP matrix such that $M = N^2$ (respectively, $P = Q^2$).

Knowing that the dimension of $\mathbb{R}^{n \times p}$ as a vector space on $\mathbb{R}$ is np, any $np \times np$ symmetric definite positive matrix T defines an inner product on $\mathbb{R}^{n \times p}$, denoted $\langle ., . \rangle_{\mathrm{T}}$ and defined component-wise as

$$\langle A , B \rangle_{\mathrm{T}} = \sum_{ij,k\ell} T_{ij,k\ell} \, a_{ij} b_{k\ell}.$$

Then, $\mathbb{R}^{n \times p}$ is endowed with a Euclidean structure, which induces a norm, which induces a metric structure with

$$d_{\mathrm{T}}(A, B) = \| A - B \|_{\mathrm{T}},$$

with

$$\| A - B \|_{\mathrm{T}}^2 = \langle A - B , A - B \rangle_{\mathrm{T}}.$$

However, we will not deal with general inner products (i.e., general SDP matrices T), but with some that can be built in a way consistent with inner products in $\mathbb{R}^n$ and $\mathbb{R}^p$. In fact, the columns live in $\mathbb{R}^n$, representing variables, and the rows live in $\mathbb{R}^p$, representing individuals. The correlation between variables, or distances between individuals, is a natural pattern in data analysis. We show how a Euclidean structure in $\mathbb{R}^n$ and a second one in $\mathbb{R}^p$ can be naturally extended to a Euclidean structure on $\mathbb{R}^{n \times p}$. This will encompass several classical developments of PCA, like PCA with weights on columns, or, as will be seen in Chap. 5, Correspondence Analysis. Centered-scaled PCA can be read this way too.

To see this, let us use tensor-based notations, i.e., considering $\mathbb{R}^{n \times p} \simeq \mathbb{R}^n \otimes \mathbb{R}^p$ (see Appendix B.4). Therefore, let $a, x \in \mathbb{R}^n$ and $b, y \in \mathbb{R}^p$. Let $T = Z^2 \in \mathbb{R}^{np \times np}$ be an inner product on $\mathbb{R}^n \otimes \mathbb{R}^p$. We wish to have on elementary (= rank one) matrices

$$\langle a \otimes b , x \otimes y \rangle_T = \langle a, x \rangle_N \langle b, y \rangle_P,$$

which extends

$$\langle a \otimes b , x \otimes y \rangle = \langle a, x \rangle \langle b, y \rangle$$

in canonical inner products. We have by definition

$$\begin{cases} \langle a \otimes b , x \otimes y \rangle_T = \langle Z(a \otimes b) , Z(x \otimes y) \rangle & \text{in } \mathbb{R}^{n \times p} \\ \langle a, x \rangle_N \quad\quad = \langle Ma, Mx \rangle & \text{in } \mathbb{R}^n \\ \langle b, y \rangle_P \quad\quad = \langle Qb, Qy \rangle & \text{in } \mathbb{R}^p . \end{cases} \tag{4.1.1}$$

So

$$\begin{aligned} \langle Z(a \otimes b) , Z(x \otimes y) \rangle &= \langle Ma, Mx \rangle \langle Qb, Qy \rangle \\ &= \langle Ma \otimes Qb , Mx \otimes Qy \rangle, \end{aligned} \tag{4.1.2}$$

which is fulfilled by selecting

$$\begin{aligned} Z(a \otimes b) &= Ma \otimes Qb \\ &= M(a \otimes b)Q. \end{aligned} \tag{4.1.3}$$

Such a property can be extended by linearity to $\mathbb{R}^{n \times p}$, yielding for any matrix $A \in \mathbb{R}^{n \times p}$

$$ZA = MAQ, \quad \text{with} \quad \begin{cases} Z \quad\quad\quad \in \mathbb{R}^{np \times np} \\ A, MAQ \quad \in \mathbb{R}^{n \times p} \simeq \mathbb{R}^{np} . \end{cases} \tag{4.1.4}$$

It defines Z, and hence $T = Z^2$. It is easy to see that T is an SDP matrix. Indeed, $T(A, B) = \langle MAQ, MBQ \rangle$. So, $T(A, A) = \langle MAQ, MAQ \rangle = \|MAQ\|^2 \geq 0$. If $T(A, A) = 0$, then $\|MAQ\| = 0$, so $MAQ = 0$ and as M, Q are invertible, $A = 0$. Finally, $T(A, B) = T(B, A)$.

This Euclidean structure is denoted

$$T \simeq N \otimes P \quad\quad \text{or} \quad\quad Z \simeq M \otimes Q$$

by a slight abuse of notation. Indeed, we have

$$T(a \otimes b) = Na \otimes Pb \neq (N \otimes P)(a \otimes b).$$

We should rigorously derive here an operator, denoted, for example, $\boxtimes$, and defined by

$$(N \boxtimes P)(a \otimes b) = Ma \otimes Pb.$$

For the sake of simplicity, we adopt the abuse of notations $T \simeq N \otimes P$ when it specifies the SDP matrix T on $\mathbb{R}^{n \times p}$ built from SDP matrix N on $\mathbb{R}^n$ and P on $\mathbb{R}^p$. So, we have

$$\|A\|_{N \otimes P} = \|MAQ\|. \tag{4.1.5}$$

If $(\mathbb{R}^{n \times p}, N \otimes P)$ denotes $\mathbb{R}^{n \times p}$ embedded with metrics $N \otimes P$ and $(\mathbb{R}^{n \times p}, \mathbb{I}_{np})$ with a standard Euclidean structure, this defines an isometry $\mathcal{I}$ by

$$(\mathbb{R}^{n \times p}, N \otimes P) \xrightarrow{\ \mathcal{I}\ } (\mathbb{R}^{n \times p}, \mathbb{I}_{np})$$

$$A \longrightarrow A' = MAQ,$$

i.e.,

$$\|\mathcal{I}(A)\| = \|A\|_{N \otimes P}.$$

We have also

$$\mathcal{I}^{-1}(A') = M^{-1}A'Q^{-1}.$$

This will permit to transport all results of Chap. 2 on PCA with a standard Euclidean structure to matrices in spaces embedded with any Euclidean structure. Here, we focus on the ones consistent with Euclidean structures in $\mathbb{R}^n$ and $\mathbb{R}^p$.

4.2   Setting the Problem

To see this, we start from the algebraic approach of PCA:

$$\left|\begin{array}{ll} \text{Given} & \text{a matrix } A \in \mathbb{R}^n \otimes \mathbb{R}^p \\ & \text{a rank } r < p \\[2ex] \text{find} & \text{a matrix } A_r \in \mathbb{R}^n \otimes \mathbb{R}^p \text{ of rank } r \\[2ex] \text{such that } \|A - A_r\| \text{ is minimum} \end{array}\right.$$

to set the problem of PCA with metrics on rows and columns as

Given	a matrix	$A \in \mathbb{R}^{n \times p}$
	an inner product in $\mathbb{R}^n$ defined by	$N \in \mathbb{R}^{n \times n}$
	an inner product in $\mathbb{R}^p$ defined by	$P \in \mathbb{R}^{p \times p}$
	with	$N = M^2, \quad P = Q^2$
	a rank	$0 < r < p$
Define	the inner product T on $\mathbb{R}^{n \times p}$	$T = Z^2$
	associated with (N, P)	$ZA = MAQ$
Find	a matrix	$A_r \in \mathbb{R}^{n \times p}$
with		rank $A_r = r$
such that		$\|A - A_r\|_T \quad$ is minimal.

4.3 Solving the Problem

A matrix A_r of rank r can be written as

$$A_r = \sum_{i=1}^{r} y_i \otimes v_i, \qquad \text{with} \quad y_i \in \mathbb{R}^n, \quad v_i \in \mathbb{R}^p,$$

with $(v_i)_i$ being an orthonormal family

$$\langle v_i , v_j \rangle = \delta_i^j,$$

which is the decomposition of matrix A on basis $(v_i)_i$. We have $y_i = A_r v_i$ (see Appendix B.4). Then

$$\|A - A_r\| = \left\| A - \sum_{i=1}^{r} y_i \otimes v_i \right\|.$$

If $A = U \Sigma V^{\mathrm{T}}$ is the SVD of A, the solution of PCA (finding A_r such that $\|A - A_r\|$ is minimal) is given by y_i that is column i of $Y = U \Sigma$ and v_i is column i of V.

Let us select the inner product $T \simeq N \otimes P$ on $\mathbb{R}^{n \times p}$ as

$$\langle A, B \rangle_{N \otimes P} = \langle MAQ, MBQ \rangle.$$

Then

$$\|A\|_{N \otimes P} = \|MAQ\|.$$

We are looking for matrix A_r such that $\|A - A_r\|_{N \otimes P}$ is minimal. We recall (see Appendix B.4) that $M(y \otimes v)Q = My \otimes Q^T v = My \otimes Qv$ as Q is symmetric. Then

$$\|A - A_r\|_{N \otimes P} = \left\| M \left(A - \sum_{i=1}^{r} y_i \otimes v_i \right) Q \right\|$$

$$= \left\| MAQ - \sum_{i=1}^{r} My_i \otimes Qv_i \right\|.$$

(4.3.1)

In the r.h.s., one recognizes the SVD of $R = MAQ$ with a standard Euclidean structure. Indeed, $\langle Qv_i, Qv_j \rangle = \langle v_i, v_j \rangle_P = \delta_i^j$ and the $(Qv_i)_i$ form an orthonormal family for the standard Euclidean structure.

Let us denote

$$\begin{cases} My_i & = x_i \\ Qv_i & = w_i. \end{cases}$$

(4.3.2)

The problem can be formulated as

$$\left| \begin{array}{ll} \text{find} & (x_i)_i, \quad (w_i)_i \\ \text{with} & (w_i)_i \text{ an orthonormal family} \\ \text{such that} & \left\| MAQ - \sum_{i=1}^{r} x_i \otimes w_i \right\| \quad \text{minimal.} \end{array} \right.$$

Then, $\{(w_i)_i, (x_i)_i\}$ are the solution of the PCA of MAQ (principal axes and principal components). The components $(y_i)_i$ and new basis vector $(v_i)_i$ of the PCA with metrics can be recovered from Eq. (4.3.2) simply by

$$\begin{cases} v_i & = Q^{-1} w_i \\ y_i & = M^{-1} x_i. \end{cases}$$

(4.3.3)

The results can be summarized as

$$\begin{cases} A_r & = \sum_{i=1}^{r} y_i \otimes v_i \\ MAQ_r & = \sum_{i=1}^{r} x_i \otimes w_i, \end{cases} \qquad (4.3.4)$$

with

$$\begin{cases} y_i = M^{-1}x_i \\ v_i = Q^{-1}w_i, \end{cases} \qquad (4.3.5)$$

hence the algorithm:

Algorithm 6 PCA of a matrix with double metrics: PCA_MET(A, M, Q)

1: **input** $A \in \mathbb{R}^{n \times p}$; $P \in \mathbb{R}^{p \times p}$, SDP ; $N \in \mathbb{R}^{n \times n}$, SDP
2: **compute** $M = N^{1/2}, Q = P^{1/2}$
3: **compute** $R = MAQ$
4: **compute** $X, \Lambda, W = $ PCA_CORE(R)
5: **compute** $Y = M^{-1}X$
6: **compute** $V = Q^{-1}W$
7: **return** Y, Λ, V

Remark The metrics on $\mathbb{R}^n$ and $\mathbb{R}^p$ are given, respectively, by N and P, which are symmetric, definite, and positive (SDP). The matrices involved in this algorithm are, respectively, M and Q, with $M = N^{1/2}$ and $Q = P^{1/2}$. They can be computed from an SVD of N and P, respectively. As N is symmetric, its SVD reads

$$N = U\Sigma U^\mathsf{T}.$$

Then

$$M = U\Sigma^{1/2}U^\mathsf{T}.$$

Indeed, $M^2 = (U\Sigma^{1/2}U^\mathsf{T})(U\Sigma^{1/2}U^\mathsf{T}) = U\Sigma^{1/2}U^\mathsf{T}U\Sigma^{1/2}U^\mathsf{T} = U\Sigma U^\mathsf{T} = N$.

Summary   Here is a summary of the calculation:

A	$n \times p$	dataset	
N	$n \times n$	metrics on $\mathbb{R}^n$	
P	$p \times p$	metrics on $\mathbb{R}^p$	
M	$n \times n$		$M = N^{1/2}$
Q	$p \times p$		$Q = P^{1/2}$
R	$n \times p$	matrix on which to run a standard PCA	$R = MAQ$
U', Σ, W		SVD of R	$R = U'\Sigma W^{\mathrm{T}}$
X	$n \times p$	principal component of R	$X = U'\Sigma$
Y	$n \times p$	Principal components of (A, M, Q)	$Y = M^{-1}X$
Λ		eigenvalues of the PCA of (A, M, Q)	$\Lambda = \Sigma^2$
V	$p \times p$	Principal axis of (A, M, Q)	$V = Q^{-1}W$
Λ		eigenvalues of the PCA of (A, M, Q)	$\Lambda = \Sigma^2$

which can be presented with the diagram:

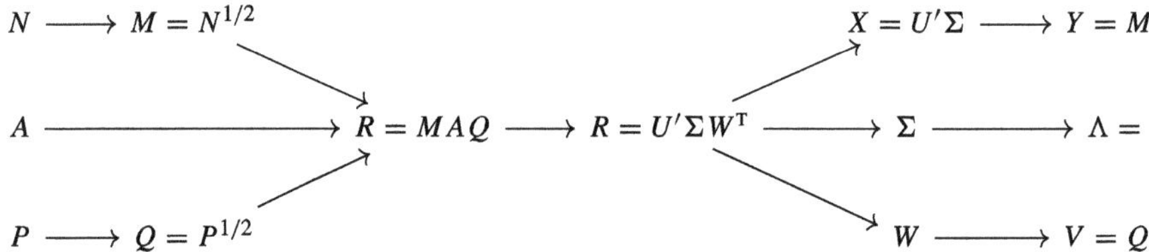

A Remark About the Calculation   Principal components $(y_i)_i$ and principal axis (v_i) are the solution of

$$\left|\begin{array}{rl} R & = MAQ \\ R^{\mathrm{T}} R w_i & = \lambda_i w_i \\ x_i & = R w_i \\ v_i & = Q^{-1} w_i \\ y_i & = M^{-1} x_i, \end{array}\right. \tag{4.3.6}$$

with

$$x_i, y_i \in \mathbb{R}^n, \qquad v_i, w_i \in \mathbb{R}^p.$$

We have

$$\left.\begin{array}{ll} R^{\mathrm{T}}R & = QA^{\mathrm{T}}NAQ \\ R^{\mathrm{T}}Rw_i = \lambda_i w_i \\ w_i & = Qv_i \end{array}\right\} \quad \Longrightarrow \quad QA^{\mathrm{T}}NAPv_i = \lambda_i Qv_i, \qquad (4.3.7)$$

and, as Q is invertible,

$$A^{\mathrm{T}}NAPv_i = \lambda_i v_i. \qquad (4.3.8)$$

This might lead to a way of computing the principal axes $(v_i)_i$ directly without computing the $(w_i)_i$ before. However, the matrix $R^{\mathrm{T}}R$ is symmetric, whereas matrix $A^{\mathrm{T}}NAP$ is not. It is known that the numerical computation of eigenvectors and eigenvalues of symmetric matrices is more accurate and robust than of nonsymmetric matrices (see Appendix B.1). Hence, it is recommended to compute first $(w_i)_i$ as solutions of $R^{\mathrm{T}}Rw_i = \lambda_i w_i$ and then the principal axis as $v_i = Q^{-1}w_i$.

4.4   Examples

Here are some classical examples of methods which can be read as PCA with weights.

Metrics as Weights

A common situation is when metrics are given in space $\mathbb{R}^p$ only, i.e., $N = \mathbb{I}_n$. Each column is a variable, measured on the same set of individuals. The matrix P formalizes a correlation structure between the variables. Then

$$MAQ = AQ,$$

and

$$\begin{aligned} R^{\mathrm{T}}R &= (MAQ)^{\mathrm{T}}(MAQ) \\ &= (AQ)^{\mathrm{T}}(AQ) \qquad (4.4.1) \\ &= Q^{\mathrm{T}}A^{\mathrm{T}}AQ. \end{aligned}$$

The most classical situation is when N is a set of weights, one per row. A weight on row i can be the confidence in the measures for the individual i. Then, $P = \mathbb{I}_p$ and

$$MAQ = MA.$$

Hence

$$R^{\mathsf{T}}R = (MAQ)^{\mathsf{T}}(MAQ)$$
$$= (MA)^{\mathsf{T}}(MA)$$
$$= A^{\mathsf{T}}MMA \qquad (4.4.2)$$
$$= A^{\mathsf{T}}NA.$$

If metrics are given on both rows and columns, we have

$$R^{\mathsf{T}}R = QA^{\mathsf{T}}NAQ. \qquad (4.4.3)$$

Geometric Approach with Weights

Let $\mathcal{A}$ be the point cloud of n points in $\mathbb{R}^p$ attached to matrix A. The distances between the points do not reflect the distances induced by the inner products (N, P). Let us denote by $\mathcal{R}$ the point cloud in $\mathbb{R}^p$ attached to matrix $R = MAQ$ (point i in $\mathcal{R}$ has as coordinates the coordinates of row i of R). If N, P are diagonal matrices with weights v_i, π_j, respectively, then

$$MAQ = [\sqrt{v_i \pi_j}\, a_{ij}]_{i,j},$$

and in $\mathbb{R}^p$

$$d^2(r_i, r_k) = \sum_j (\sqrt{v_i \pi_j}\, a_{ij} - \sqrt{v_k \pi_j}\, a_{kj})^2$$
$$= \sum_j \pi_j\, (\sqrt{v_i}\, a_{ij} - \sqrt{v_k}\, a_{kj})^2. \qquad (4.4.4)$$

This is the distance between points of the point cloud $\mathcal{A}_{\mathrm{N}}$ in $\mathbb{R}^p$ attached to matrix MA with the inner product in $\mathbb{R}^p$ defined by weight matrix P for computing distances. This will be useful for Correspondence Analysis (see Chap. 5).

Scaled PCA

An apparently straightforward and standard application of PCA with metrics is scaled PCA. Let $A \in \mathbb{R}^{n \times p}$ be a column-wise centered matrix, i.e.,

$$\sum_i a_i = 0. \qquad (4.4.5)$$

The variances (or norms) of columns of A can vary significantly. In such a case, the correlation matrix $C = A^{\mathrm{T}} A$ can be dominated by rows and columns corresponding to the variable with the largest variance. Scaled PCA is clipping this uninteresting result off, by giving equal weights to each variable. The technical trick is to equalize variances between columns, by dividing each column j by its standard deviation (or norm). If $a_{*j} \in \mathbb{R}^n$ is column j of A, this reads

$$a_{*j} \longrightarrow a'_{*j} = \frac{a_{*j}}{\|a_{*j}\|}.$$

(4.4.6)

Hence

$$\|a'_{*j}\| = 1.$$

(4.4.7)

This can be read as a PCA with inner product $N = \mathbb{I}_n$ in $\mathbb{R}^n$ and $P = \mathrm{diag}\left(\frac{1}{\|a_{*j}\|}\right)$ in $\mathbb{R}^p$. So $MAQ = A'$, and PCA of A' is run. However, it is not exactly a PCA with metrics as presented in this chapter, because PCA is run on A', or $\mathcal{I}(A)$, and the result is not the PCA of A with metrics given by P.

Notes and References   The problem (and solution) of PCA with weights on rows and columns can be found in [Rao64] or [Gre84]. It is presented in [Jol02, sect. 14.2]. The algebraic approach with generalization to metrics in Euclidean spaces has been proposed as a general method with the notion of duality diagram in [CP79] which has been at the root of many works (see [PCY79]). The formalism presented here with tensor product can be found in [Fra92].

4.5   Classical Analysis of a Matrix with Metrics and Weights: A Geometric Approach

The geometric school of linear dimension reduction has developed a framework for analyzing a matrix $A \in \mathbb{R}^{n \times p}$ with a metric defined by P in $\mathbb{R}^p$ and weights $(w_i)_i$ on the individuals. The point cloud is made with rows $(a_i)_i$ of A with $a_i \in \mathbb{R}^p$. It is presented here and will be very useful for understanding the geometric approach of CoA.

- **Setting the problem:** Rows of A (points of the point cloud) are in $\mathbb{R}^p$. We set the problem for the best projection of the point cloud on a one-dimensional space in $\mathbb{R}^p$ spanned by a vector v with $\|v\|_{\mathrm{P}} = 1$. The aim is to compute v. The projection of a_i on $\mathbb{R}v$ is the vector $\lambda v \in \mathbb{R}v$ such that $d_{\mathrm{P}}(a_i, \lambda v) = \|a_i - \lambda v\|_{\mathrm{P}}$ is minimal. It is given by (see Sect. B.2)

$$\mathcal{P}_v a_i = \langle a_i, v \rangle_{\mathrm{P}} v = \langle P a_i, v \rangle v,$$

and its norm is

$$\|\mathcal{P}_v a_i\|_P^2 = \langle Pa_i, v\rangle^2.$$

Let

$$\mathcal{J} = \sum_{i=1}^{n} \|\mathcal{P}_v a_i\|_P^2 = \sum_{i=1}^{n} \langle Pa_i, v\rangle^2$$

be the inertia of the point cloud attached to A for the inner product P (it is denoted by $\mathcal{J}$ and not $\mathcal{I}$ to avoid a confusion with isometry $\mathcal{I}$). The first step without weights on rows but with metrics defined by P is to find v with $\|v\|_P = 1$ such that $\mathcal{J}$ is maximal. The final step is to introduce the weights and define

$$\mathcal{J} = \sum_{i=1}^{n} w_i \langle Pa_i, v\rangle^2$$

as the inertia of the point cloud with inner product defined by P and weights on rows defined by w. Then, the geometrical approach can be set as

given	a matrix $A \in \mathbb{R}^{n \times p}$
with	row i denoted $a_i \in \mathbb{R}^p$
an inner product	$P \in \mathbb{R}^{p \times p}$
a set of row weights	$w \in \mathbb{R}^n$
find	a vector $v \in \mathbb{R}^p$
with	$\|v\|_P = 1$
such that	$\mathcal{J} = \sum_{i=1}^{n} w_i \|\mathcal{P}_v a_i\|^2$ is maximal
where	$\mathcal{P}_v$ is the projection on $\mathbb{R}v$ in $\mathbb{R}^p$.

- **Solving the problem:** Let us note first that

$$\|\mathcal{P}_v a_i\|^2 = \langle Pa_i, v\rangle^2. \tag{4.5.1}$$

Then

$$\mathcal{J} = \sum_{i} w_i \langle Pa_i, v\rangle^2 = \sum_{i} \langle \sqrt{w_i}\, P\, a_i\,,\, v\rangle^2. \tag{4.5.2}$$

Let us observe that:

→ The vectors $Pa_i \in \mathbb{R}^p$ are the row vectors of matrix AP (indeed, P is symmetric by definition).

→ The terms $\sqrt{w_i}\,Pa_i$ are the rows of matrix $D_w^{1/2}AP$ if D_w is the $n \times n$ diagonal matrix with w in the diagonal ($D_w[i,i] = w_i, \quad i \neq j \implies D[i,j] = 0$).

Then $\mathcal{J}$ can be rewritten as

$$\mathcal{J} = \|D_w^{1/2}APv\|^2. \tag{4.5.3}$$

So, the problem can be restated as

$$
\begin{aligned}
&\text{given} &&A, P, w \text{ as above,}\\
&\text{find} &&v \in \mathbb{R}^p\\
&\text{with} &&\|v\|_P^2 = 1\\
&\text{such that} &&\mathcal{J} = \|D_w^{1/2}APv\|^2 \text{ is maximal.}
\end{aligned}
$$

To solve this with Lagrange multipliers (an optimum with constraints), let us denote

$$H = D_w^{1/2}AP. \tag{4.5.4}$$

Then, $\mathcal{J} = \|Hv\|^2$, and the problem can be set as maximizing $\mathcal{J} = \|Hv\|^2$ knowing that $\|Qv\|^2 = 1$. We have

$$\frac{\partial \mathcal{J}}{\partial v} = 2H^{\mathsf{T}}H, \qquad \frac{\partial \|v\|_P^2}{\partial v} = 2Pv, \tag{4.5.5}$$

so there is a scalar $\lambda \in \mathbb{R}$ such that

$$H^{\mathsf{T}}Hv = \lambda Pv. \tag{4.5.6}$$

This can be written

$$PA^{\mathsf{T}}D_w^{1/2}D_w^{1/2}APv = PA^{\mathsf{T}}D_wAPv = \lambda Pv, \tag{4.5.7}$$

and Pv is an eigenvector of $PA^{\mathsf{T}}D_wA$.

One may be tempted to note that, as P is invertible (it is an SDP matrix), one has

$$A^{\mathsf{T}}D_wAPv = \lambda v, \tag{4.5.8}$$

and v is an eigenvector of $A^{\mathrm{T}} D_w A P$. However, $A^{\mathrm{T}} D_w A P$ is not symmetric, and it is preferable for stability and robustness of the result to keep as a result the solution of Eq. (4.5.7)

$$P A^{\mathrm{T}} D_w A P v = \lambda P v. \tag{4.5.9}$$

Let us define v' by $v' = Q v$, or $P v = Q v'$. Then

$$P A^{\mathrm{T}} D_w A Q v' = \lambda Q v'. \tag{4.5.10}$$

Multiplying both members of this equation on the left by Q^{-1} yields

$$Q A^{\mathrm{T}} D_w A Q v' = \lambda v', \tag{4.5.11}$$

and v' is an eigenvector of symmetric matrix $Q A^{\mathrm{T}} D_w A Q$. Then, it is easy to recover v by $v = Q^{-1} v'$.

Notes and References This section is adapted from [LMP00, 1.1.6] where it is presented as a diversification of the general analysis (PCA). Similar approaches can be found in [LMT77, LMF82].

Chapter 5
Correspondence Analysis

Abstract Correspondence Analysis can be presented in several equivalent ways. In a traditional setting, it is the analysis of a contingency table, i.e., a cross-table between two qualitative variables. Two point clouds are associated with such a table: one where one point is a row of the table, i.e., a profile for one modality of the variable attached to the rows, and a second where one point is a column, i.e., one profile attached to the second variable. Each point cloud is given a series of weights and equipped with a distance, in such a way that both point clouds can be projected simultaneously to reveal the dependency structure between the variables, or rows and columns, hence the name Correspondence Analysis, as a correspondence between both point clouds. In this chapter, we start with the observation that Correspondence Analysis is the analysis of the contingency table with two metrics on row and column space each related to weights chosen as inverse to the margins of the table. It is shown that the norm of the table is the χ^2 norm and that Correspondence Analysis is the task of finding the table of low rank as close as possible to the table with χ^2 norm. The equivalence between both approaches, and the correspondence between them, is presented.

A remarkable application of the PCA with weights on rows and columns is the development of Correspondence Analysis as the analysis of a contingency table with metrics associated with its margins.

Let us adopt here some standard notations for contingency tables. A contingency table T is a table of counts of n items allocated to categories of two variables. Indices of the values of the first variable are denoted i and of the second variable j. The value n_{ij} in row i and column j of T is the number of items in category i for the first variable and j for the second. It is standard to denote that $i \in [\![1, I]\!]$ and $j \in [\![1, J]\!]$. Then

$$T \in \mathbb{R}^{I \times J} \simeq \mathbb{R}^I \otimes \mathbb{R}^J.$$

A. Franc, *Linear Dimensionality Reduction*, Lecture Notes in Statistics 228,
https://doi.org/10.1007/978-3-031-95785-7_5

It is standard to denote

$$n_{i+} = \sum_{j} n_{ij}, \quad n_{+j} = \sum_{i} n_{ij}, \qquad n_{++} = \sum_{i,j} n_{ij} = \sum_{i} n_{i+} = \sum_{j} n_{+j}.$$

5.1 Link with χ^2 Distance

We first establish a link between the norm of a contingency table with metrics associated with margins on rows and columns on the one hand and the χ^2 statistics of the table on the other. Let us recall that the χ^2 statistics permits to derive a test to decide whether rows and columns are independent or not. The χ^2 can be considered as a variance of the contingency table (it will be made rigorous below) or a measure of the discrepancy between the observed table and the virtual table built by assuming independence between rows and columns with observed row and column margins.

Let us denote

$$A = \frac{T}{n_{++}}. \tag{5.1.1}$$

The general term in A is denoted a_{ij}, and we have

$$A \in \mathbb{R}^{I \times J} \simeq \mathbb{R}^{I} \otimes \mathbb{R}^{J}. \tag{5.1.2}$$

Let us denote, respectively, by $r \in \mathbb{R}^{I}$ and $c \in \mathbb{R}^{J}$ the marginal sums of A on rows and columns:

$$\begin{cases} r_i = \sum_{j} a_{ij} = a_{i+} \\ c_j = \sum_{i} a_{ij} = a_{+j}. \end{cases} \tag{5.1.3}$$

Let us denote by D_r and D_c the square diagonal matrices with diagonal, respectively, r and c:

$$D_r = \operatorname{diag} r, \qquad D_c = \operatorname{diag} c. \tag{5.1.4}$$

In case of independence between both variables, the expectation for A is

$$\widetilde{A} = r \otimes c. \tag{5.1.5}$$

Indeed, we have, ignoring the value of the other variable, if X_r is the random variable for row marginals (i.e., an item is in row i) and X_c for column marginals (i.e., an item is in column j),

$$P(X_r = i) = r_i, \qquad P(X_c = j) = c_j.$$

Then, in case of independence, the probability for an item to be in row i and in column j is

$$P(X_r = i \,;\, X_c = j) = r_i c_j.$$

Then, a first observation is that (see Chap. 4 for notation $\|.\|_{D_r^{-1} \otimes D_c^{-1}}$)

$$\chi^2(A) = \left\| A - \widetilde{A} \right\|_{D_r^{-1} \otimes D_c^{-1}}^2. \tag{5.1.6}$$

Proof Indeed, we have

$$\chi^2(A) = \sum_{i,j} \frac{(a_{ij} - r_i c_j)^2}{r_i c_j}, \qquad \text{by definition}$$

$$= \sum_{i,j} \left(\frac{a_{ij} - r_i c_j}{\sqrt{r_i c_j}} \right)^2$$

$$= \sum_{i,j} \left(\frac{1}{\sqrt{r_i}} (a_{ij} - r_i c_j) \frac{1}{\sqrt{c_j}} \right)^2 \tag{5.1.7}$$

$$= \left\| D_r^{-1/2} (A - r \otimes c) D_c^{-1/2} \right\|^2$$

$$= \| A - r \otimes c \|_{D_r^{-1} \otimes D_c^{-1}}^2.$$

$\square$

5.2   Description of the Method

Then, Correspondence Analysis is a partition of the variance $\| A - r \otimes c \|_{D_r^{-1} \otimes D_c^{-1}}$ concentrated on the first axes. It is henceforth a **PCA** of $A - r \otimes c$ with metrics defined by D_r^{-1} on rows and D_c^{-1} on columns, i.e., with weights $1/r_i$ on row i and $1/c_j$ on column j.

$$
\begin{array}{ll}
\text{given} & T = (n_{ij})_{i,j} \text{ (a contingency table)} \\[2ex]
\text{compute} & n_{++} = \sum_{i,j} n_{ij} \\
 & A = \dfrac{T}{n_{++}} \\
 & r_i = \sum_j a_{ij} \\
 & c_j = \sum_i a_{ij} \\[2ex]
\text{run} & \texttt{PCAmet} \\
\text{on} & A - r \otimes c \\[2ex]
\text{with diagonal metrics} & 1/r_i \text{ on row } i \\
 & 1/c_j \text{ on column } j
\end{array}
$$

We have

$$
\begin{cases}
M = \operatorname{diag} 1/\sqrt{r_i} = D_r^{-1/2} \\
Q = \operatorname{diag} 1/\sqrt{c_j} = D_c^{-1/2}.
\end{cases}
$$

Then

$$
A_{\mathrm{M} \otimes \mathrm{Q}} = M(A - r \otimes c)Q
$$
$$
= \left[\frac{a_{ij} - r_i c_j}{\sqrt{r_i c_j}} \right]_{i,j}. \tag{5.2.1}
$$

This yields the following algorithm, which is a pure transposition of Algorithm 6:

Algorithm 7 Correspondence Analysis of a contingency table: $\mathrm{COA}(T)$

1: **input** $T \in \mathbb{R}^{I \times J}$, a contingency table, of general term n_{ij}
2: **compute** $A = T/n_{++}$,   with $n_{++} = \sum_{i,j} n_{ij}$
3: **compute** $r_i = \sum_j a_{ij}$, $D_r = \operatorname{diag} r$
4: **compute** $c_j = \sum_i a_{ij}$, $D_c = \operatorname{diag} c$
5: **compute** $M = \operatorname{diag}(1/\sqrt{r_i})$
6: **compute** $Q = \operatorname{diag}(1/\sqrt{c_j})$
7: **compute** $A_{\mathrm{M,Q}} = M(A - r \otimes c)Q = \left[\frac{a_{ij} - r_i c_j}{\sqrt{r_i c_j}} \right]_{i,j}$
8: **compute** $Z, \Lambda, X = \mathrm{PCA\_CORE}(A_{\mathrm{M,Q}})$
9: **compute** $Y_r = M^{-1} Z$
10: **compute** $Y_c = Q^{-1} X$
11: **return** Y_r, Y_c, Λ

5.3 CoA and Geometry of Point Clouds

Let us consider the matrix $X \in \mathbb{R}^{I \times J}$ with elements

$$x_{ij} = \frac{a_{ij}}{\sqrt{r_i c_j}}. \tag{5.3.1}$$

Its PCA is the PCAmet of A with inner product defined by D_r^{-1} on rows and D_c^{-1} on columns. We show here that its PCA has a tight link with the CoA of A.

Let us define for this calculation vectors denoted $\sqrt{r} \in \mathbb{R}^I$ and $\sqrt{c} \in \mathbb{R}^J$ by

$$\sqrt{r} = (\sqrt{r_1}, \ldots, \sqrt{r_I}), \qquad \sqrt{c} = (\sqrt{c_1}, \ldots, \sqrt{c_J})$$

and show that

$$X^{\mathsf{T}} X \sqrt{c} = \sqrt{c}.$$

Therefore, we show that $X\sqrt{c} = \sqrt{r}$. Indeed, for row i of $X\sqrt{c}$

$$\left(X\sqrt{c}\right)_i = \sum_j \frac{a_{ij}\sqrt{c_j}}{\sqrt{r_i c_j}} = \frac{1}{\sqrt{r_i}} \sum_j a_{ij} = \frac{r_i}{\sqrt{r_i}} = \sqrt{r_i}.$$

Then, we show that $X^{\mathsf{T}}\sqrt{r} = \sqrt{c}$. So, $X^{\mathsf{T}} X \sqrt{c} = \sqrt{c}$. Indeed, for row j of $X^{\mathsf{T}}\sqrt{r}$

$$\left(X^{\mathsf{T}}\sqrt{r}\right)_j = \sum_i \frac{a_{ij}}{\sqrt{r_i c_j}} \sqrt{r_i} = \frac{1}{\sqrt{c_j}} \sum_i a_{ij} = \frac{c_j}{\sqrt{c_j}} = \sqrt{c_j}.$$

So, $X^{\mathsf{T}} X \sqrt{c} = X^{\mathsf{T}}\sqrt{r} = \sqrt{c}$, and $\sqrt{c}$ is an eigenvector of $X^{\mathsf{T}} X$ associated with the eigenvalue $\lambda = 1$, i.e., a principal axis of X. The associated principal component is $X\sqrt{c} = \sqrt{r}$. This triplet $(\sqrt{r}, \sqrt{c}, 1)$ is trivial (i.e., it exists whatever the matrix A) and brings the information that r and c are the vectors of margins of A. The full SVD of X is

$$X = \sqrt{r} \otimes \sqrt{c} + \sum_{j=2}^{J} \sigma_j y_j \otimes v_j$$

(indeed, $\sigma_1 = 1$, $\|\sqrt{r}\|^2 = \sum_i r_i = 1$, and $\|\sqrt{c}\| = \sum_j c_j = 1$). Then, $\sum_{j=2}^{J} \sigma_j y_j \otimes v_j$ is the SVD of $X - \sqrt{r} \otimes \sqrt{c} = D_r^{-1/2} A D_c^{-1/2}$, i.e., the CoA of A.

We can build two point clouds from X, one for the rows, $\mathcal{X}$, and one for the columns, $\mathcal{X}'$.

$\diamond$ $\mathcal{X}$ is a cloud of I points $x_i \in R^J$ of coordinates

$$x_i = (x_{ij})_{1 \le j \le J} = (x_{i1}, \ldots, x_{iJ}). \tag{5.3.2}$$

If $\mathbb{R}^J$ is equipped with standard inner product (i.e., $\mathbb{I}_J$) and if x_i, x_k are two points in $\mathcal{X}$, we have

$$
\begin{aligned}
d(x_i, x_k)^2 &= \sum_j \left(\frac{a_{ij}}{\sqrt{r_i c_j}} - \frac{a_{kj}}{\sqrt{r_k c_j}} \right)^2 \\
&= \sum_j \frac{1}{c_j} \left(\frac{a_{ij}}{\sqrt{r_i}} - \frac{a_{kj}}{\sqrt{r_k}} \right)^2.
\end{aligned}
\tag{5.3.3}
$$

$\diamond$ $\mathcal{X}'$ is a cloud of J points $x'_j \in R^I$ of coordinates

$$x'_j = (x_{ij})_{1 \le i \le I} = (x_{1j}, \ldots, x_{Ij}). \tag{5.3.4}$$

If $\mathbb{R}^I$ is equipped with standard inner product (i.e., $\mathbb{I}_I$) and if x'_j, x'_ℓ are two points in $\mathcal{X}'$, we have

$$
\begin{aligned}
d(x'_j, x'_\ell)^2 &= \sum_i \left(\frac{a_{ij}}{\sqrt{r_i c_j}} - \frac{a_{i\ell}}{\sqrt{r_i c_\ell}} \right)^2 \\
&= \sum_i \frac{1}{r_i} \left(\frac{a_{ij}}{\sqrt{c_j}} - \frac{a_{i\ell}}{\sqrt{c_\ell}} \right)^2.
\end{aligned}
\tag{5.3.5}
$$

This leads to a classical presentation of Correspondence Analysis as a correspondence between two point clouds, as presented below. We have not denoted $\mathcal{R}$ but $\mathcal{X}$ the point cloud of the rows of X, nor $\mathcal{C}$ but $\mathcal{X}'$ the point cloud of its columns, because $\mathcal{X} \ne \mathcal{R}$ and $\mathcal{X}' \ne \mathcal{C}$. $\mathcal{R}$ and $\mathcal{C}$ are defined below.

5.4 Classical Presentation

There is a classical presentation of Correspondence Analysis as an analysis of two point clouds associated with a contingency table, one for the rows and one for the columns (which play a symmetric role), each with weights and metrics. This is the geometric approach. It emphasizes that two point clouds, and not simply one, can be built: one for rows and one for columns, and CoA can be seen as a simultaneous analysis of both. CoA establishes a *correspondence* between both point clouds.

Let us recall that if T is a contingency table, F (the matrix of frequencies) is built from T by dividing it by the sum of all its elements:

$$T = (n_{ij})_{i,j} \longrightarrow n_{++} = \sum_{i,j} n_{ij} \longrightarrow F : f_{ij} = \frac{n_{ij}}{n_{++}}.$$

To comply with standard notations in classical textbooks (see below), we denote by F, and not A, the matrix of frequencies. We will denote

For row		For columns
$f_{i+} = \sum_j f_{ij}$	and	$f_{+j} = \sum_i f_{ij}$
$f_{i*} = (f_{i1}, \ldots, f_{iJ}) \in \mathbb{R}^J$	and	$f_{*j} = (f_{1j}, \ldots, f_{Ij}) \in \mathbb{R}^I$
$r = (f_{1*}, \ldots, f_{I*}) \in R^I$	and	$c = (f_{*1}, \ldots, f_{*J}) \in R^J$

Two point clouds are classically attached to F:

- A cloud $\mathcal{R}$ of row profiles, as I points $(x_i)_i$ in $\mathbb{R}^J$, with point x_i of coordinates

$$x_i = \left[\frac{f_{ij}}{f_{i+}} \right]_j \in \mathbb{R}^J. \tag{5.4.1}$$

- A cloud $\mathcal{C}$ of column profiles, as J points y_j in $\mathbb{R}^I$, with point y_j of coordinates

$$y_j = \left[\frac{f_{ij}}{f_{+j}} \right]_i \in \mathbb{R}^I. \tag{5.4.2}$$

Then, metrics with diagonal matrices, respectively, D_c^{-1} for $\mathbb{R}^J$ and D_r^{-1} for $\mathbb{R}^I$ are selected, with D_c being the $J \times J$ diagonal matrix of term f_{+j} and D_r the $I \times I$ diagonal matrix of term f_{i+}. Hence, distances between points x_i and x_k in $\mathbb{R}^J$ are computed as

$$d^2(x_i, x_k) = \sum_j \frac{1}{f_{+j}} \left(\frac{f_{ij}}{f_{i+}} - \frac{f_{kj}}{f_{k+}} \right)^2 \tag{5.4.3}$$

and between points y_j and y_ℓ in $\mathbb{R}^{\mathrm{I}}$ as

$$d^2(y_j, y_\ell) = \sum_i \frac{1}{f_{i+}} \left(\frac{f_{ij}}{f_{+j}} - \frac{f_{i\ell}}{f_{+\ell}} \right)^2 . \tag{5.4.4}$$

These weights tend to give an equal importance to all modalities of a variable, whatever their size. It is analogous to weighting by the inverse of the variance in scaled PCA. A further property often invoked is that if two categories of the same variable have the same profile (say, i and i' for which $\forall j,\ f_{ij}/f_{i+} = f_{i'j}/f_{i'+}$), then it is logical to lump them together into a single category, and this must not modify the distances between row profiles.

It is then standard to define both point clouds as (see [Gre84, sect. 4.1], [LMF82, sect. IV.5.], [Sap90, sect. 10.1], and [LMP00, sect. 1.3.3])

	$\mathcal{R}$: row profiles in $\mathbb{R}^J$	$\mathcal{C}$: column profiles in $\mathbb{R}^I$
Point cloud	$R = D_r^{-1} F$	$C = D_c^{-1} F^{\mathrm{T}}$
Metric	D_c^{-1}	D_r^{-1}
Weights	r	c

It is easy to get lost in the indices, the rows, the columns, and spaces $\mathbb{R}^I$ and $\mathbb{R}^J$. A row profile is in $\mathbb{R}^J$, and its coefficients are indexed by j; a column profile is in $\mathbb{R}^I$, and its coefficients are indexed by i. With coefficients, this yields

$$R_{ij} = \frac{f_{ij}}{f_{i+}}, \qquad C_{ij} = \frac{f_{ij}}{f_{+j}} . \tag{5.4.5}$$

The barycenters $g^{(\mathrm{R})} \in \mathbb{R}^J$ of R and $g^{(\mathrm{C})} \in \mathbb{R}^I$ of C are, respectively, c and r (and not r and c). Indeed,

$$g_j^{(\mathrm{R})} = \sum_i f_{i+} \frac{f_{ij}}{f_{i+}} = \sum_i f_{ij} = f_{+j} = c_j , \tag{5.4.6}$$

and

$$g_i^{(\mathrm{C})} = \sum_j f_{+j} \frac{f_{ij}}{f_{+j}} = \sum_j f_{ij} = f_{i+} = r_i . \tag{5.4.7}$$

Centering column-wise the point cloud $\mathcal{R}$ is subtracting $g_i^{(C)} = c$ to each row, hence making the transformation

$$R \longrightarrow R - \mathbf{1}_I \otimes c,$$

where $\mathbf{1}_I \otimes c$ is the matrix in $\mathbb{R}^{I \times J}$ with c on each row. Then, the inertia I_R of centered point cloud $R - \mathbf{1}_I \otimes c$ with weights r on rows and metric D_c^{-1} in $\mathbb{R}^J$ is (explained step by step)

$$
\begin{aligned}
I_R^2 &= \sum_i r_i \, \| R_{i*} - c \|_{D_c^{-1}}^2 & R_{i*} &= (R_{i1}, \ldots, R_{iJ}) \\[2mm]
&= \sum_i f_{i+} \left\| \frac{f_{i*}}{f_{i+}} - c \right\|_{D_c^{-1}}^2, & R_{i*} &= \frac{f_{i*}}{f_{i+}}, \; r_i = f_{i+} \\[2mm]
&= \sum_i f_{i+} \left(\frac{1}{f_{+j}} \sum_j \left(\frac{f_{ij}}{f_{i+}} - f_{+j} \right)^2 \right), & c_j &= f_{+j}, \; D_c^{-1} = \operatorname{diag}\left(\frac{1}{f_{+j}} \right) \\[2mm]
&= \sum_i f_{i+} \left(\sum_j \left(\frac{f_{ij}}{f_{i+}\sqrt{f_{+j}}} - \sqrt{f_{+j}} \right)^2 \right) & \sqrt{f_{+j}} &= \frac{f_{+j}}{\sqrt{f_{+j}}} \\[2mm]
&= \sum_{i,j} \left(\sqrt{f_{i+}} \frac{f_{ij}}{f_{i+}\sqrt{f_{+j}}} - \sqrt{f_{i+}}\sqrt{f_{+j}} \right)^2 \\[2mm]
&= \sum_{i,j} \left(\frac{f_{ij} - f_{i+}f_{+j}}{\sqrt{f_{i+}f_{+j}}} \right)^2,
\end{aligned}
$$

$$(5.4.8)$$

where we recognize the χ^2 norm of $F = (f_{ij})_{i,j}$ or the norm of F with metrics defined by D_c^{-1} in $\mathbb{R}^J$ (space of rows) and by D_r^{-1} in $\mathbb{R}^I$ (space of columns). Its partition with concentration of the inertia on the first components is the CoA of F.

If I_C^2 is the inertia of row-wise centered point cloud C in $\mathbb{R}^I$, i.e., of $C - \mathbf{1}_p \otimes r$ with weights c on rows and metric D_r^{-1} in $\mathbb{R}^I$, a similar calculation yields

$$
I_C^2 = \sum_{i,j} \left(\frac{f_{ij} - f_{i+}f_{+j}}{\sqrt{f_{i+}f_{+j}}} \right)^2 = I_R^2. \tag{5.4.9}
$$

Both inertia are equal, and the analysis of both point clouds is a geometric guise of the CoA of F.

5.5 Geometric Approach

To see this, one can use the developments presented in Sect. 4.5. Let us first present a translation of the notations between Sect. 4.5 and here:

PCAmet with weight		CoA on R	CoA on C
A	$\longleftrightarrow$	$R = D_r^{-1}F$	$C = D_c^{-1}F^{\mathrm{T}}$
P	$\longleftrightarrow$	D_c^{-1}	D_r^{-1}
w	$\longleftrightarrow$	r	c

Let us recall that if

$$H = D_w^{1/2}AP, \tag{5.5.1}$$

the first principal axis u is the solution of

$$H^{\mathrm{T}}Hu = \lambda\, Pu, \tag{5.5.2}$$

with

$$H^{\mathrm{T}}Hu = PA^{\mathrm{T}}D_wAPu = \lambda Pu, \tag{5.5.3}$$

and, as P is invertible as an SDP matrix, u is the solution of

$$A^{\mathrm{T}}D_wAPu = \lambda u. \tag{5.5.4}$$

We then have, for CoA on row profiles,

$$\rightarrow \quad H = D_r^{1/2}D_r^{-1}FD_c^{-1} = D_r^{-1/2}FD_c^{-1},$$
$$\rightarrow \quad H^{\mathrm{T}}H = \left(D_c^{-1}F^{\mathrm{T}}D_r^{-1/2}\right)\left(D_r^{-1/2}FD_c^{-1}\right) = D_c^{-1}F^{\mathrm{T}}D_r^{-1}FD_c^{-1},$$

and, after simplification by D_c^{-1}, u is the solution of

$$F^{\mathrm{T}}D_r^{-1}FD_c^{-1}u = \lambda u. \tag{5.5.5}$$

Similarly, for CoA on column profiles,

$$\rightarrow \quad H = D_c^{1/2}D_c^{-1}F^{\mathrm{T}}D_r^{-1} = D_c^{-1/2}F^{\mathrm{T}}D_r^{-1},$$
$$\rightarrow \quad H^{\mathrm{T}}H = \left(D_r^{-1}FD_c^{-1/2}\right)\left(D_c^{-1/2}F^{\mathrm{T}}D_r^{-1}\right) = D_r^{-1}FD_c^{-1}F^{\mathrm{T}}D_r^{-1},$$

and, after simplification by D_r^{-1}, the first axis v is the solution of

$$FD_c^{-1}F^{\mathrm{T}}D_r^{-1}v = \lambda v \tag{5.5.6}$$

(see [LMP00, p. 83], with, here again, a translation of notations).

If we multiply Eq. (5.5.5) on the left by D_c^{-1} and set $u' = D_c^{-1}u$, we have

$$D_c^{-1}F^{\mathrm{T}}D_r^{-1}FD_c^{-1}u = \lambda D_c^{-1}u, \tag{5.5.7}$$

or

$$CRu' = \lambda u'. \tag{5.5.8}$$

Similarly, multiplying Eq. (5.5.6) left by D_r^{-1} and setting $v' = D_r^{-1}v$ yield

$$RCv' = \lambda v'. \tag{5.5.9}$$

This yields

$$\begin{cases} v' = Ru' \\ u' = Cv', \end{cases} \tag{5.5.10}$$

or

$$\begin{cases} v = D_r^{-1}RD_c^{-1}u \\ u = D_c^{-1}CD_r^{-1}v. \end{cases} \tag{5.5.11}$$

Notes and References Correspondence Analysis has a long history and has been object of long debates, renaming, and rediscoveries between different schools. Correspondence Analysis has been proposed first by Hirshfeld in 1935 (Hirschfeld, M. O., 1935, A connection between correlation and contingency. *Proc. Camb. Phil. Soc.*, **31**:520–524). It has been rediscovered by Guttman in 1959 (Guttman, L., 1959, Metricizing rank ordered and unordered data for a linear factor analysis. *Sankhyā*, **21**:257–268). The link between CoA and reciprocal averaging has been presented in [Hil74]. Correspondence Analysis has been rediscovered and studied independently by several researchers, such as J.-P. Benzecri, in 1962, in the context of mathematical linguistics inspired by the works of Chomsky, J. de Leeuw in Netherlands, and C. Hayashi in Japan. His type III Quantification methods, published in the 1950s (Hayashi, C. (1954). Multidimensional quantification with applications to analysis of social phenomena. *Annals of the Institute of Statistical Mathematics*, **5(2)**:121–143.), is equivalent to Correspondence Analysis. A historical background and synthesis is given in [TY85]. One of the early works in the French school is

Cordier, B., Sur l'analyse Factorielle des Correspondances. *PhD, Rennes*, 1965. This approach has been developed by Greenacre in Pretoria who has studied with J.-P. Benzecri [Gre84]. The algorithm presented here is the one presented in [NG07]. We have used also [Sap90, chapter 10] and [LMP00, sect. 1.3] for the geometric interpretation.

Chapter 6
PCA with Instrumental Variables

Abstract Up to now, we have seen PCA and PCA with metrics on rows and columns. A matrix A being given, PCA is about finding new axes and components such that the quality of representation of A by a matrix of low prescribed rank by components on its first new axis is best. PCA with metrics is about solving this problem with any Euclidean distance in spaces spanned by rows or columns of A. PCA with instrumental variables is about solving this problem when linear constraints are given on the axes and components, i.e., they live in spaces of lower dimension. The constraints on columns, for example, can be given by a second matrix, denoted B, with as many rows as A, such that the principal components of A must live in the space spanned by the columns of B. It is shown that a solution is given by first projecting the given matrix on the space spanned by the constraints on rows and columns, e.g., projecting the columns of A on the space spanned by the columns of B, and then performing the PCA of the projected matrix. This leads to three observations: (i) Finding the first axis is equivalent to finding a linear combination of the columns of B with the best correlation with all columns of A, i.e., finding the best linear regression of the columns of B which explain all columns of A, (ii) PCA with instrumental variables is equivalent to PLS, and (iii) it is possible to combine PCA with metrics and PCA with instrumental variables, which leads to CoA with instrumental variables.

Here, we use tensor notation for PCA (the reader not familiar with tensor notation can consult Appendix B.4). Let $y \in \mathbb{R}^n$ and $v \in \mathbb{R}^p$. For the sake of simplicity and as a "crash course" for these notations, we can accept as a definition that $\otimes$ is a bilinear map

$$\mathbb{R}^n \times \mathbb{R}^p \xrightarrow{\;\otimes\;} \mathbb{R}^{n \times p}$$

$$(y, v) \longrightarrow y \otimes v,$$

A. Franc, *Linear Dimensionality Reduction*, Lecture Notes in Statistics 228, https://doi.org/10.1007/978-3-031-95785-7_6

defined by

$$y \otimes v := yv^{\mathrm{T}}.$$

It can be illustrated by

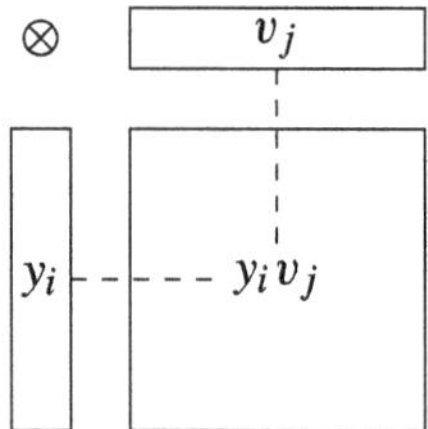

if $y = (y_1, \ldots, y_n) \in \mathbb{R}^n$ and $v = (v_1, \ldots, v_p) \in \mathbb{R}^p$. Then $y \otimes v \in \mathbb{R}^{n \times p}$ and

$$(y \otimes v)_{ij} = y_i v_j.$$

$\mathbb{R}^{n \times p}$ is denoted as well $\mathbb{R}^n \otimes \mathbb{R}^p$.

Let $A \in \mathbb{R}^{n \times p}$. The best rank-$r$ approximation of A is denoted $A_r = Y_r V_r^{\mathrm{T}}$, with $A_r \in \mathbb{R}^{n \times p}$, $Y_r \in \mathbb{R}^{n \times r}$, and $V_r \in \mathbb{R}^{p \times r}$. Then

$$A_r = \sum_{a=1}^{r} y_a v_a^{\mathrm{T}}, \qquad y_a = A v_a,$$

with $(y_a)_a$ and $(v_a)_a$ being the columns of Y and V, respectively, ordered in decreasing order of singular values of A. Equivalently

$$A_r = \sum_{a} y_a \otimes v_a, \qquad \text{with} \begin{cases} y_a & \in \mathbb{R}^n \\ v_a & \in \mathbb{R}^p. \end{cases}$$

PCA with instrumental variables (PCAiv) is setting some constraints on the principal components $(y_a)_a$ or principal axes $(v_a)_a$, i.e., that they live in given subspaces, respectively, $F \subset \mathbb{R}^n$ and $H \subset \mathbb{R}^p$.

Usually, those spaces are given as spanned by a set of vectors in, respectively, $\mathbb{R}^n$ (for F, constraint on y) and $\mathbb{R}^p$ (for H, constraints on v). A classical situation is when there is no constraint on the axis v, but only on the components y, and F is spanned by an $n \times q$ matrix denoted I, the columns of which are called *instrumental variables*.

6.1 Setting the Problem

Let us suppose that $\dim F = m$ and $\dim H = q$. PCAiv can be stated as

$$
\begin{aligned}
&\text{given} && A \in \mathbb{R}^{n \times p} \\
& && F \subset \mathbb{R}^n, \quad H \subset \mathbb{R}^p \\
& && \text{with } \dim F = m, \ \dim H = q \\
& && 0 < r < \min(m, q) \\[1em]
&\text{find} && A_r \in F \otimes H \\[1em]
&\text{such that} && \|A - A_r\| \text{ is minimal}
\end{aligned}
$$

This problem can be set by selecting a basis of F and one of H. We assume here that they are orthonormal. If they are not orthonormal, it is possible to build an orthonormal base by QR decomposition, for example, with the Gram-Schmidt orthogonalization procedure (which can be numerically unstable but is easy to implement), or using Householder reflections (more stable).

6.2 Solving the Problem

As $F \subset \mathbb{R}^n$ and $H \subset \mathbb{R}^p$, $F \otimes H \subset \mathbb{R}^n \otimes \mathbb{R}^p$, where $\subset$ means "is a vector subspace of."

- **Main result:** Let $A_{F \otimes H}$ be the projection of A on $F \otimes H$. Then, the solution A_r of PCAiv is the PCA of $A_{F \otimes H}$.

Proof Let us denote by $\mathcal{P}$ the projection from $\mathbb{R}^n \otimes \mathbb{R}^p$ on $F \otimes H$ and by $\mathcal{P}^\perp$ the projection on $(F \otimes H)^\perp$. Let us recall that $A_r = \sum_j y_j \otimes v_j$. If $\mathbb{I}$ is the identity in $\mathbb{R}^n \otimes \mathbb{R}^p$, we have $\mathbb{I} = \mathcal{P} + \mathcal{P}^\perp$ and

$$
\begin{aligned}
\|A - A_r\|^2 &= \left\| \left(\mathcal{P} + \mathcal{P}^\perp \right) (A - A_r) \right\|^2 && \text{as } \mathbb{I} = \mathcal{P} + \mathcal{P}^\perp \\
&= \left\| \mathcal{P}(A - A_r) + \mathcal{P}^\perp (A - A_r) \right\|^2 \\
&= \left\| \mathcal{P}(A - A_r) \right\|^2 + \left\| \mathcal{P}^\perp (A - A_r) \right\|^2 && \text{by Pythagoras} \\
&= \left\| \mathcal{P}(A - A_r) \right\|^2 + \left\| \mathcal{P}^\perp (A) \right\|^2 && \text{as } \mathcal{P}^\perp A_r = 0 \\
&= \left\| \mathcal{P}(A) - A_r \right\|^2 + \left\| \mathcal{P}^\perp (A) \right\|^2 && \text{as } \mathcal{P} A_r = A_r.
\end{aligned}
$$

Let us recall that A, F, H are fixed, hence so are $\mathcal{P}, \mathcal{P}^\perp$. Then, A_r only can vary. Hence $\|A - A_r\|^2$ is minimum when $\|\mathcal{P}(A) - A_r\|^2$ is minimum, and A_r is the PCA of $\mathcal{P}(A) = A_{F \otimes H}$.

Before developing the main result, we need to recall some elementary results on linear projectors. Let $F \subset \mathbb{R}^n$, and denote $\mathcal{P}_F$ the projector on F. Let $U_F = (u_1, \ldots, u_m)$ be an orthonormal basis of F column-wise. Then

$$\mathcal{P}_F = U_F U_F^T. \tag{6.2.1}$$

Indeed, if $x \in \mathbb{R}^n$, we have

$$\mathcal{P}_F x = \sum_i \langle u_i, x \rangle u_i$$

$$= \sum_i (u_i \otimes u_i) x$$

$$= \left(\sum_i u_i \otimes u_i \right) x.$$

Then

$$\mathcal{P}_F = \sum_i u_i \otimes u_i = U_F U_F^T. \tag{6.2.2}$$

Let $\mathcal{P}_F$ be the projector on F, $\mathcal{P}_H$ be the projector on H, and $A \in \mathbb{R}^{n \times p}$. Then, the projector $\mathcal{P}_{F \otimes H}$ on $F \otimes H$ is defined by

$$\mathcal{P}_{F \otimes H} A = U_F U_F^T A V_H V_H^T. \tag{6.2.3}$$

Indeed, if $A = \sum_{a=1}^p y_a \otimes v_a$,

$$A_{F \otimes H} = \sum_{a=1}^p \mathcal{P}_F y_a \otimes \mathcal{P}_H v_a$$

$$= \sum_{a=1}^p P_F (y_a \otimes v_a) P_H \qquad (P_H \text{ is symmetric})$$

$$= P_F \left(\sum_{a=1}^p y_a \otimes v_a \right) P_H$$

$$= P_F A P_H,$$

hence the algorithm:

Algorithm 8 Naive pseudocode for $\texttt{PCAiv}(A, U_{\mathrm{F}}, V_{\mathrm{H}})$

1: **input:** $A \in \mathbb{R}^{n \times p}$
2: **input:** U_{F} orthonormal, $F = \operatorname{span} U_{\mathrm{F}} \subset \mathbb{R}^{n}$,
3: **input:** V_{H} orthonormal, $H = \operatorname{span} V_{\mathrm{H}} \subset \mathbb{R}^{p}$,
4: **compute** $P_{\mathrm{F}} = U_{\mathrm{F}} U_{\mathrm{F}}^{\mathrm{T}}$
5: **compute** $P_{\mathrm{H}} = V_{\mathrm{H}} V_{\mathrm{H}}^{\mathrm{T}}$
6: **compute** $T = P_{\mathrm{F}} A P_{\mathrm{H}}$
7: **compute** $(Y, \Lambda, V) = \textsc{pca\_core}(T)$
8: **return** Y, Λ, V

- **Algorithm:** In this algorithm, the SVD is run in the PCA of $T = P_{\mathrm{F}} A P_{\mathrm{H}} \in \mathbb{R}^{n \times p}$, the same dimensions as A. Its complexity is in $\mathcal{O}(n^2 p)$. However, as $T \in F \otimes H$, with $\dim F = m$ and $\dim H = q$, the rank of T is $q < p$ (if we assume that $q \leq m$). Here, we show how to take the low rank into account to lower the complexity of the SVD, i.e., to run it on a matrix of dimension $m \times q$, i.e., with complexity in $\mathcal{O}(m^2 q)$. Therefore, we start from

$$
\begin{aligned}
T &= P_{\mathrm{F}} A P_{\mathrm{H}} \\
&= (U_{\mathrm{F}} U_{\mathrm{F}}^{\mathrm{T}}) A (V_{\mathrm{H}} V_{\mathrm{H}}^{\mathrm{T}}) \\
&= U_{\mathrm{F}} (U_{\mathrm{F}}^{\mathrm{T}} A V_{\mathrm{H}}) V_{\mathrm{H}}^{\mathrm{T}} \\
&= U_{\mathrm{F}} T' V_{\mathrm{H}}^{\mathrm{T}},
\end{aligned}
$$

denoting

$$
T' = U_{\mathrm{F}}^{\mathrm{T}} A V_{\mathrm{H}} \qquad \in \mathbb{R}^{m \times q}.
$$

Let $(U_{\mathrm{T}}', \Sigma_{\mathrm{T}}', V_{\mathrm{T}}')$ be the SVD of T'

$$
T' = U_{\mathrm{T}}' \Sigma_{\mathrm{T}}' V_{\mathrm{T}}'^{\mathrm{T}}, \tag{6.2.4}
$$

with (this will be useful soon), assuming $m \geq q$ here for the sake of simplicity

$$
\begin{cases}
U_{\mathrm{F}} & \in \mathbb{R}^{n \times m} \\
U_{\mathrm{T}}' & \in \mathbb{R}^{m \times q} \\
\Sigma_{\mathrm{T}}' & \in \mathbb{R}^{q \times q} \\
V_{\mathrm{T}}' & \in \mathbb{R}^{q \times q} \\
V_{\mathrm{H}} & \in \mathbb{R}^{p \times q}.
\end{cases}
$$

Then

$$T = U_{\mathrm{F}} U'_{\mathrm{T}} \Sigma'_{\mathrm{T}} V'_{\mathrm{T}} V_{\mathrm{H}}^{\mathrm{T}}$$
$$= (U_{\mathrm{F}} U'_{\mathrm{T}}) \Sigma'_{\mathrm{T}} (V_{\mathrm{H}} V'_{\mathrm{T}})^{\mathrm{T}}, \tag{6.2.5}$$

and $(U_{\mathrm{F}} U'_{\mathrm{T}},\ \Sigma'_{\mathrm{T}},\ V'^{\mathrm{T}}_{\mathrm{T}} V^{\mathrm{T}}_{\mathrm{H}})$ is the SVD of T because

$$\begin{cases} (U_{\mathrm{F}} U'_{\mathrm{T}})^{\mathrm{T}}(U_{\mathrm{F}} U'_{\mathrm{T}}) = U'^{\mathrm{T}}_{\mathrm{T}} U^{\mathrm{T}}_{\mathrm{F}} U_{\mathrm{F}} U'_{\mathrm{T}} \\ \qquad\qquad\qquad\quad = U'^{\mathrm{T}}_{\mathrm{T}} \mathbb{I}_m U'_{\mathrm{T}} \\ \qquad\qquad\qquad\quad = U'^{\mathrm{T}}_{\mathrm{T}} U'_{\mathrm{T}} \\ \qquad\qquad\qquad\quad = \mathbb{I}_q, \end{cases}$$

so $U_{\mathrm{F}} U'_{\mathrm{T}}$ is orthonormal. Similarly

$$\begin{cases} (V_{\mathrm{H}} V'_{\mathrm{T}})^{\mathrm{T}}(V_{\mathrm{H}} V'_{\mathrm{T}}) = V'^{\mathrm{T}}_{\mathrm{T}} V^{\mathrm{T}}_{\mathrm{H}} V_{\mathrm{H}} V'_{\mathrm{T}} \\ \qquad\qquad\qquad\quad = V'^{\mathrm{T}}_{\mathrm{T}} \mathbb{I}_q V'_{\mathrm{T}} \\ \qquad\qquad\qquad\quad = V'^{\mathrm{T}}_{\mathrm{T}} V'_{\mathrm{T}} \\ \qquad\qquad\qquad\quad = \mathbb{I}_q, \end{cases}$$

and $V_{\mathrm{H}} V'^{\mathrm{T}}_{\mathrm{T}}$ is orthonormal too. This small calculation replaces the SVD of $T \in \mathbb{R}^{n \times p}$ in $\mathcal{O}(n^2 p)$ by the SVD of $T' \in \mathbb{R}^{m \times q}$ in $\mathcal{O}(m^2 q)$. This leads to the following algorithm:

Algorithm 9 Pseudocode for $\mathtt{PCAiv}(A,\ U_{\mathrm{F}},\ V_{\mathrm{H}})$

1: **input:** $A \in \mathbb{R}^{n \times p}$
2: **input:** $F = \mathrm{span}\ U_{\mathrm{F}} \subset \mathbb{R}^n$, U_{F} orthonormal
3: **input:** $H = \mathrm{span}\ V_{\mathrm{H}} \subset \mathbb{R}^p$, V_{H} orthonormal
4: **compute** $T' = U^{\mathrm{T}}_{\mathrm{F}} A V_{\mathrm{H}}$
5: **do** the SVD of T': $T' = U'_{\mathrm{T}} \Sigma'_{\mathrm{T}} V'^{\mathrm{T}}_{\mathrm{T}}$
6: **compute** $U = U_{\mathrm{F}} U'_{\mathrm{T}}$
7: **compute** $V = V_{\mathrm{H}} V'_{\mathrm{T}}$
8: **compute** $\Lambda = \Sigma'^{2}_{\mathrm{T}}$
9: **compute** $Y = U \Sigma'_{\mathrm{T}}$
10: **return** Y, Λ, V

Summary Here is a summary of the calculation:

A	$n \times p$	dataset	
U_F	$n \times m$	orthonormal matrix spanning $F \subset R^n$	
V_H	$p \times q$	orthonormal matrix spanning $H \subset R^p$	
T'	$m \times q$	auxiliary matrix for calculations	$T' = U_\mathrm{F}^\mathrm{T} A V_\mathrm{H}$
$U_\mathrm{T}', \Sigma_\mathrm{T}', V_\mathrm{T}'$		SVD of T'	$T' = U_\mathrm{T}' \Sigma_\mathrm{T}' V_\mathrm{T}'^\mathrm{T}$
U	$n \times q$		$U = U_\mathrm{F} U_\mathrm{T}'$
Y	$n \times q$	principal components	$Y = U \Sigma_\mathrm{T}'$
Λ	$q \times q$	eigenvalues	$\Lambda = \Sigma_\mathrm{T}'^2$
V	$p \times q$	principal axis	$V = V_\mathrm{H} V_\mathrm{T}'$

which can be presented with the diagram:

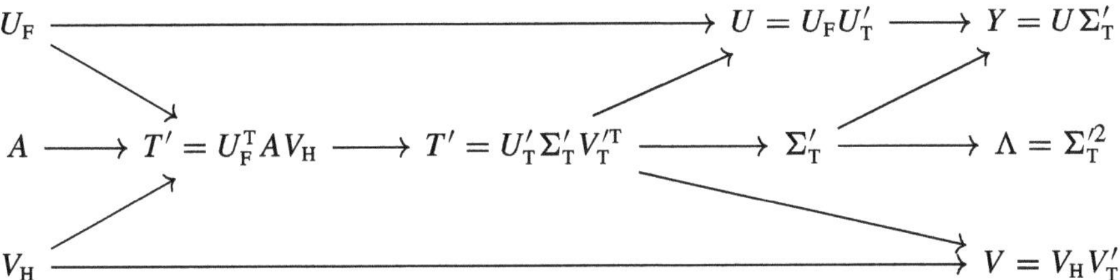

6.3   Interpretation of PCAiv

`PCAiv` is a two-step procedure:

1. Project the variables into the space spanned by the instrumental variables.
2. Run the `PCA` of the projected variables.

This can be sketched by the following diagram:

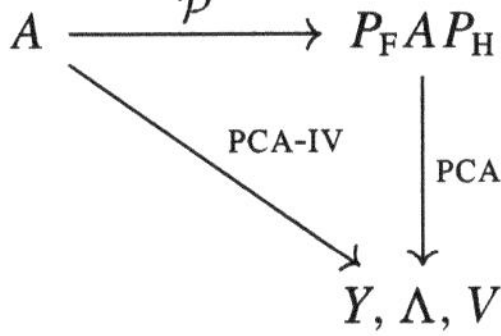

or

$$\text{PCA-IV}(A) = \text{PCA}\,(\mathcal{P}(A)).$$

This leads to:

1. Another interpretation of PCAiv
2. An estimate of the quality of the PCAiv

as follows.

Let us give an interpretation of PCAiv on the example of a constraint on the principal components only. Let $a \in \mathbb{R}^n$. The projection of a on F is given by $P_{\mathrm{F}}a$. Let now $A \in \mathbb{R}^{n \times p}$. The matrix $\widetilde{A} = P_{\mathrm{F}}\,A$ is in $\mathbb{R}^{n \times p}$, like A, and its column j is the projection of column j of A on F. It is the regression of the column j by the basis vectors of F, given by the columns of U_{F}. The first principal component of $\widetilde{A}$ is the best summary of A by one single linear combination of the columns of U_{F}. It is the best same linear regression applied to all columns of A. In this interpretation, PCAiv is equivalent to PLS.

For the PCA, classical estimators of the quality of the PCA can be used. Let us denote by Y_r the matrix of r first principal components of $P_{\mathrm{F}}AP_{\mathrm{H}}$ on which PCA is run, and

$$\|Y_r\| = \rho_r \|P_{\mathrm{F}}AP_{\mathrm{H}}\|, \tag{6.3.1}$$

i.e., the quality of the PCA is denoted by ρ_r at rank r. The quality of the projection can be quantified by

$$\|P_{\mathrm{F}}AP_{\mathrm{H}}\| = \theta\,\|A\|. \tag{6.3.2}$$

(θ is the cosine of the angle between A and $P_{\mathrm{F}}AP_{\mathrm{H}}$). We then have

$$\|Y_r\| = \rho_r\theta\|A\|, \tag{6.3.3}$$

and the quality of the PCAiv can be poor for two reasons:

$\rightarrow$ The quality θ of the projection is poor.
$\rightarrow$ The quality ρ_r of the PCA at rank r is poor.

It is essential to distinguish between the quality of the projection and the quality of the PCA. Let us see it on an example. We assume that the matrix A is built as a low rank matrix plus some important noise. It can be expected that the noise is poorly projected (there is no specific subspace where the noise is better represented), whereas there is some specific low-dimensional subspace where the low rank component of A is well represented. Then, projection will filter out the noise, whereas the PCA of the projected matrix will find the low rank property of the structure of A. θ will be low, but ρ_r close to 1. Even if the quality $\rho_r\theta$ is poor because of the poor quality θ of the projection, PCAiv is a success as it has filtered out the noise and exhibited the low rank of the dataset.

6.4 Non-orthonormal Basis

In Sect. 6.1, the subspaces F and H are given by their basis U_F and V_H, respectively. Let us now suppose that F and H are, respectively, spanned by the columns of U'_F and V'_H, which are no longer assumed to be orthonormal. Then

$$P_F = U'_F (U'^T_F U'_F)^{-1} U'^T_F, \qquad P_H = V'_H (V'^T_H V'_H)^{-1} V'^T_H,$$

and Eq. (6.2.3) reads

$$T = P_F A P_H$$
$$= U'_F (U'^T_F U'_F)^{-1} U'^T_F \ A \ V'_H (V'^T_H V'_H)^{-1} V'^T_H.$$

However, this equation, even if correct, is not efficient for numerical analysis where it should be avoided because of the cost of three products and one inversion per projector. It is far more efficient to work with an orthogonal basis of F and H, by:

1. Building an orthonormal basis U_F of F and an orthonormal basis V_H of H by QR decomposition or the Gram-Schmidt orthonormalization
2. Building the projectors $P_F = U_F U^T_F$ and $P_H = V_H V^T_H$

6.5 PCA with Metrics and Instrumental Variables

Those methods, PCAmet and PCAiv, can be associated like pieces of puzzle to build a chain of treatments.

PCAiv is running the PCA of A with constraints on principal axes and components which must belong to subspaces of, respectively, $\mathbb{R}^p$ and $\mathbb{R}^n$:

$$\begin{cases} y_j \in F \subset \mathbb{R}^n \\ v_j \in H \subset \mathbb{R}^p. \end{cases} \tag{6.5.1}$$

If U (respectively, V) is an orthonormal matrix with a basis of E (respectively, F) as column vectors, this is done by building the projectors

$$\mathbb{R}^n \xrightarrow{R=UU^T} F, \qquad \mathbb{R}^p \xrightarrow{S=VV^T} H \tag{6.5.2}$$

and running the PCA of $A' = RAS$, the projection of A on $F \otimes H$.

If $\mathbb{R}^n$ and $\mathbb{R}^p$ are endowed with metrics given by N and P, respectively, running the PCA of A' with those metrics is running the PCA of MAQ with $M = N^{1/2}$ and $Q = P^{1/2}$. However, the projectors R and S depend on the metrics N and P.

Let us write the projector on $H \subset \mathbb{R}^p$ with the inner product defined by P first. Let $x \in \mathbb{R}^p$ and $v \in H$ with $\|v\|_P = 1$. The projection x' of x on $H = \mathbb{R}v$ is given by

$$
\begin{aligned}
x' &= \langle x, v \rangle_P v \\
&= \langle x, Pv \rangle v \\
&= \langle Pv, x \rangle v \\
&= (v \otimes Pv).x.
\end{aligned}
\tag{6.5.3}
$$

Hence, the projector on $\mathbb{R}v$ with the inner product defined by P is $v \otimes Pv$. If $H = \mathrm{span}\,(v_1, \ldots, v_r)$, we have

$$
\begin{aligned}
x' &= \sum_a \langle x, Pv_a \rangle v_a \\
&= \sum_a (v_a \otimes Pv_a)x,
\end{aligned}
\tag{6.5.4}
$$

and the projector is

$$
\begin{aligned}
S &= \sum_a v_a \otimes Pv_a \\
&= V(PV)^{\mathrm{T}} \\
&= VV^{\mathrm{T}}P \\
&= PVV^{\mathrm{T}}
\end{aligned}
\tag{6.5.5}
$$

if $V \in \mathbb{R}^{p \times r}$ is the matrix with v_a in column a. The last equality comes from the observation that P and VV^{T} are symmetric; hence $(VV^{\mathrm{T}})P$ is symmetric and $VV^{\mathrm{T}}P = (VV^{\mathrm{T}}P)^{\mathrm{T}} = PVV^{\mathrm{T}}$.

Similarly, we have

$$
R = UU^{\mathrm{T}}N
\tag{6.5.6}
$$

and

$$
\begin{aligned}
A' &= RAS \\
&= UU^{\mathrm{T}}NAPVV^{\mathrm{T}}.
\end{aligned}
\tag{6.5.7}
$$

Let us now write the PCA of A' with inner products defined by N on $\mathbb{R}^n$ and P on $\mathbb{R}^p$. It is the PCA of

$$
A'' = MA'Q, \qquad \text{with} \quad N = M^2, \quad P = Q^2.
$$

Notes and References Apparently, the term *PCA with Instrumental Variables* appeared first in [Rao64]. It has been studied with a double set of constraints in [Sab84]. PCAiv is equivalent to PLS. It has been widely used in ecology (see [LSBB91]). For example, in plant ecology, when you have n observation plots, you often have two arrays: one, denoted A, with the floristic composition of each plot (plots are in row and plant species in columns), and the other, denoted B, with measurements of environmental variables in these same plots, such as soil characteristics, climate, etc. The diversity of floristic composition has many sources, like landscape history, environmental factors, and random fluctuations. PCAiv of the pair of arrays (A, B) gives two results: (i) by the projection, the fraction of the variance of the floristic distribution between plots explained by the environmental variables and (ii), by PCA, the contributions of each environmental variables to the structure of the floristic distribution. Conversely, it is possible to run PCAiv of (B, A), i.e., PCA of B with A as instrumental variable. Projection tells how much the floristic structure can be reconstructed from the environmental structure, and PCA of the projection tells which and how environmental variables enable to reconstruct this projection. This is bioindication.

Chapter 7
Canonical Correlation Analysis

Abstract This chapter looks at the correlation between two arrays. If we have two arrays, each containing the values for its own set of variables on the same set of individuals, a legitimate question is whether or not the information provided by these two sets of data is redundant. One answer is provided by Canonical Correlation Analysis, developed in this chapter. The basic idea is to extend the notion of correlation between two variables to a correlation between two arrays. The two sets of variables are considered to be correlated if each variable in one array is a linear combination of the variables in the other array. We will therefore look for axes for each of the two arrays such that the components associated with them are the most correlated. A geometric solution using projectors is proposed. However, while this result is very intuitive, it is not optimal for the complexity of the calculations. The proposed algorithm for calculating the canonical axes and components is based on a reorganization of the calculations to reduce them to a PCA with metrics of the cross-tabulated array of the two initial arrays. This approach will be used in the next chapter to extend the canonical analysis to more than two tables.

Let us have two datasets as two sets of variables (or features) on the same set of items, A and B, with $A \in \mathbb{R}^{n \times p}$ and $B \in \mathbb{R}^{n \times q}$. We assume that $p, q < n$ and even $q \leq p < n$, and $p + q \leq n$. The set of columns of each matrix spans a subspace in $\mathbb{R}^n$. If a column of A belongs to the space spanned by the columns of B, then there exists a linear regression on the columns of B which explains this column of A, and both sets of columns are correlated in $\mathbb{R}^n$. Canonical Correlation Analysis (CCA) is about finding sets of vectors ($=$ components) in the spaces spanned by the columns of each matrix with the greatest correlation. In this chapter, the problem is stated (Sect. 7.1) and solved (Sect. 7.2) first in an algebraic way, and a second effort must be done to implement involved linear algebra calculations with a reasonable cost (Sect. 7.3).

A. Franc, *Linear Dimensionality Reduction*, Lecture Notes in Statistics 228,
https://doi.org/10.1007/978-3-031-95785-7_7

7.1 Stating the Problem

We will denote by v_A a vector in $\mathbb{R}^p$ and by v_B a vector in $\mathbb{R}^q$. A vector $y_A \in$ span A (respectively, $y_B \in$ span B) can be written as Av_A (respectively, Bv_B). The correlation between y_A and y_B is

$$\mathrm{corr}\,(y_A, y_B) = \frac{\langle y_A, y_B \rangle}{\|y_A\| \|y_B\|}.$$

Then, Canonical Correlation Analysis of (A, B) for the first canonical components can be stated as

> Given $\quad A \in \mathbb{R}^{n \times p},\, B \in \mathbb{R}^{n \times q}$
>
> Find $\quad v_A \in \mathbb{R}^p,\, v_B \in \mathbb{R}^q$
>
> such that $\quad \dfrac{\langle Av_A, Bv_B \rangle}{\|Av_A\| \|Bv_B\|}$ is maximal.

As such the problem is difficult to solve. One reason is that $\|v_A\|$ and $\|v_B\|$ can take any nonzero value (the correlation remains unchanged by a rescaling of $\|v_A\|$ or $\|v_B\|$). One could add a constraint like $\|v_A\| = \|v_B\| = 1$, but the problem is still difficult to solve. There is an equivalent formulation leading to easier calculations for the solution, by setting a constraint on $y_A = Av_A$ (respectively, $y_B = Bv_B$):

> Given $\quad A \in \mathbb{R}^{n \times p},\, B \in \mathbb{R}^{n \times q}$
>
> Find $\quad v_A \in \mathbb{R}^p,\, v_B \in \mathbb{R}^q$
> with $\quad \|Av_A\|^2 = 1,\, \|Bv_B\|^2 = 1$
>
> such that $\quad \langle Av_A, Bv_B \rangle$ is maximal.

7.2   Solving the Problem

This is an optimization problem with constraints, which can be solved by Lagrange multipliers. Let us recall that if

$$\mathbb{R}^n \xrightarrow{\ f,g\ } \mathbb{R},$$

an optimum of $f(x)$ under the constraint $g(x) = 0$ is obtained at some points x satisfying

$$\nabla f - \lambda \nabla g = \mathbf{0},$$

where

$$\nabla f = \left(\frac{\partial f}{\partial x_1}, \ldots, \frac{\partial f}{\partial x_n} \right).$$

It remains to check that such a solution is a maximum (it can be a minimum or a saddle point).

Here, the unknowns are (v_A, v_B), the function f is $f(v_A, v_B) = \langle Av_A, Bv_B \rangle$ and g is $\|Av_A\|^2 = \|Bv_B\|^2 = 1$. One computes separately the partial derivatives with respect to v_A and v_B, denoting them ∇_{v_A} and ∇_{v_B}. One has

$$\begin{cases} \nabla_{v_A} \langle Av_A, Bv_B \rangle = A^{\mathsf{T}} Bv_B \quad \nabla_{v_B} \langle Av_A, Bv_B \rangle = B^{\mathsf{T}} Av_A \\ \nabla_{v_A} \|Av_A\|^2 \quad\ \ = 2A^{\mathsf{T}} Av_A \ \ \nabla_{v_B} \|Bv_B\|^2 \quad\ \ = 2B^{\mathsf{T}} Bv_B. \end{cases} \tag{7.2.1}$$

Then, the solution satisfies to

$$\begin{cases} \nabla_{v_A} : \ A^{\mathsf{T}} Bv_B = \lambda\, A^{\mathsf{T}} Av_A \\ \nabla_{v_B} : \ B^{\mathsf{T}} Av_A = \mu\, B^{\mathsf{T}} Bv_B. \end{cases} \tag{7.2.2}$$

We first show that $\lambda = \mu$.

Proof Therefore, we deduce from Eq. (7.2.2) that

$$\begin{cases} \langle A^{\mathsf{T}} Bv_B, v_A \rangle = \lambda \langle A^{\mathsf{T}} Av_A, v_A \rangle \\ \langle B^{\mathsf{T}} Av_A, v_B \rangle = \mu \langle B^{\mathsf{T}} B, v_B \rangle \end{cases}$$

and observe that

$$\begin{aligned} \langle A^{\mathsf{T}} Av_A, v_A \rangle &= \langle Av_A, Av_A \rangle = 1 \\ \langle B^{\mathsf{T}} Bv_B, v_B \rangle &= \langle Bv_B, Bv_B \rangle = 1. \end{aligned}$$

Then

$$\lambda = \langle A^{\mathrm{T}} B v_{\mathrm{B}}, v_{\mathrm{A}} \rangle = \langle B^{\mathrm{T}} A v_{\mathrm{A}}, v_{\mathrm{B}} \rangle = \mu.$$

Thus, Eq. (7.2.2) reads

$$\begin{cases} A^{\mathrm{T}} B v_{\mathrm{B}} = \lambda A^{\mathrm{T}} A v_{\mathrm{A}} \\ B^{\mathrm{T}} A v_{\mathrm{A}} = \lambda B^{\mathrm{T}} B v_{\mathrm{B}}. \end{cases} \tag{7.2.3}$$

Multiplying leftwise the first equation by $(A^{\mathrm{T}} A)^{-1}$ and the second by $(B^{\mathrm{T}} B)^{-1}$ yields

$$\begin{cases} (A^{\mathrm{T}} A)^{-1} A^{\mathrm{T}} B v_{\mathrm{B}} = \lambda v_{\mathrm{A}} \\ (B^{\mathrm{T}} B)^{-1} B^{\mathrm{T}} A v_{\mathrm{A}} = \lambda v_{\mathrm{B}}, \end{cases} \tag{7.2.4}$$

and, having in mind that $A v_{\mathrm{A}} = y_{\mathrm{A}}$ and $B v_{\mathrm{B}} = y_{\mathrm{B}}$,

$$\begin{cases} A(A^{\mathrm{T}} A)^{-1} A^{\mathrm{T}} y_{\mathrm{B}} = \lambda y_{\mathrm{A}} \\ B(B^{\mathrm{T}} B)^{-1} B^{\mathrm{T}} y_{\mathrm{A}} = \lambda y_{\mathrm{B}}. \end{cases} \tag{7.2.5}$$

One recognizes in the l.h.s. of (7.2.5)

$$\begin{cases} A(A^{\mathrm{T}} A)^{-1} A^{\mathrm{T}} = \mathcal{P}_{\mathrm{A}} \\ B(B^{\mathrm{T}} B)^{-1} B^{\mathrm{T}} = \mathcal{P}_{\mathrm{B}}, \end{cases}$$

i.e., the projectors on the spaces spanned by the columns of A and of B, respectively. This leads to

$$\begin{cases} \mathcal{P}_{\mathrm{A}} y_{\mathrm{B}} = \lambda y_{\mathrm{A}} \\ \mathcal{P}_{\mathrm{B}} y_{\mathrm{A}} = \lambda y_{\mathrm{B}} \end{cases}$$

or

$$\begin{cases} \mathcal{P}_{\mathrm{A}} \mathcal{P}_{\mathrm{B}} y_{\mathrm{A}} = \lambda^{2} y_{\mathrm{A}} \\ \mathcal{P}_{\mathrm{B}} \mathcal{P}_{\mathrm{A}} y_{\mathrm{B}} = \lambda^{2} y_{\mathrm{B}}. \end{cases} \tag{7.2.6}$$

- **Interpretation:** The interpretation is quite natural. span A and span B are two vector subspaces in $\mathbb{R}^{n}$ of dimension p and q, respectively. Let us have $y_{\mathrm{A}} \in$ span A. It is projected as $y'_{\mathrm{B}} \in$ span B by $y'_{b} = \mathcal{P}_{\mathrm{B}} y_{\mathrm{A}}$. y'_{b} itself is projected as

$y''_a \in$ span A by $y''_a = \mathcal{P}_A y'_b$. One has

$$\text{span } A \xrightarrow{\ \mathcal{P}_B\ } \text{span } B \xrightarrow{\ \mathcal{P}_A\ } \text{span } A$$

$$y_A \longrightarrow y'_B \longrightarrow y''_A.$$

The same can be written for y_B:

$$\text{span } B \xrightarrow{\ \mathcal{P}_A\ } \text{span } A \xrightarrow{\ \mathcal{P}_B\ } \text{span } B$$

$$y_B \longrightarrow y'_A \longrightarrow y''_B.$$

Equation (7.2.6) says that when the correlation between y_A and y_B is maximal, then y_A (respectively, y_b) and y''_a (respectively, y''_b) are collinear and eigenvectors of $\mathcal{P}_A \mathcal{P}_B$ (respectively, $\mathcal{P}_B \mathcal{P}_A$).

Notes and References   The computation of the solution in this section is classical and has been borrowed from [LMF82, section IV.6.4].

7.3   Computing the Solution

This solution is very intuitive, geometrically speaking, but it does not lead to the most efficient way to compute a solution. We start from

$$\begin{cases} \mathcal{P}_A \mathcal{P}_B y_A = \lambda^2 y_A \\ \mathcal{P}_B \mathcal{P}_A y_B = \lambda^2 y_B. \end{cases}$$

Let us recall that

$$\begin{cases} \mathcal{P}_A = A(A^{\mathsf{T}} A)^{-1} A^{\mathsf{T}} \\ \mathcal{P}_B = B(B^{\mathsf{T}} B)^{-1} B^{\mathsf{T}}. \end{cases}$$

Then

$$\begin{cases} A(A^{\mathsf{T}} A)^{-1} A^{\mathsf{T}} B(B^{\mathsf{T}} B)^{-1} B^{\mathsf{T}} y_A = \lambda^2 y_A \\ B(B^{\mathsf{T}} B)^{-1} B^{\mathsf{T}} A(A^{\mathsf{T}} A)^{-1} A^{\mathsf{T}} y_B = \lambda^2 y_B. \end{cases} \tag{7.3.1}$$

We show next how it is possible to decrease the complexity of computing $\mathcal{P}_A = A(A^T A)^{-1} A^T$ and $\mathcal{P}_B = B(B^T B)^{-1} B^T$ and avoid the computation of $\mathcal{P}_A \mathcal{P}_B$. Indeed, computing

$A^T A$	is in	$\mathcal{O}(np^2)$	$B^T B$	is in	$\mathcal{O}(nq^2)$
$(A^T A)^{-1}$		$\mathcal{O}(p^3)$	$(B^T B)^{-1}$		$\mathcal{O}(q^3)$
$A(A^T A)^{-1} A^T$		$\mathcal{O}(np^2)$	$B(B^T B)^{-1} B^T$		$\mathcal{O}(nq^2)$.

Hence, computing $\mathcal{P}_A$ (respectively, $\mathcal{P}_B$) is in $\mathcal{O}(np^2)$ (respectively, $\mathcal{O}(nq^2)$). Moreover, computing $\mathcal{P}_A \mathcal{P}_B$ is in $\mathcal{O}(n^3)$ as $\mathcal{P}_A, \mathcal{P}_B \in \mathbb{R}^{n \times n}$. However, rank $\mathcal{P}_A = m$ and rank $\mathcal{P}_B = q$. So, it should be possible to reduce the complexity of the calculation.

Here, we show how the calculation of the solution can avoid the complexity in $\mathcal{O}(n^3)$ of $\mathcal{P}_A \mathcal{P}_B$. We show at the same time that the solution of CCA can be read as the solution of a PCA with metrics. Without loss of generality, we assume that $p \geq q$. Let us denote

$$\begin{cases} T & = A^T B \\ N & = (A^T A)^{-1} \\ P & = (B^T B)^{-1}. \end{cases} \tag{7.3.2}$$

Then, $\mathcal{P}_A = ANA^T$, $\mathcal{P}_B = BPB^T$, and $\mathcal{P}_A \mathcal{P}_B = ANA^T BPB^T = ANTPB^T$. So,

$$\begin{cases} ANTPB^T y_A & = \lambda^2 y_A \\ BPT^T NA^T y_B & = \lambda^2 y_B. \end{cases}$$

Let us recall that

$$y_A = A v_A, \qquad y_B = B v_B.$$

Then

$$\begin{cases} ANTPT^T v_A = \lambda^2 A v_A \\ BPT^T NT v_B = \lambda^2 B v_B. \end{cases}$$

We can "simplify" by A and B by left multiplication by $(A^T A)^{-1} A^T$ and $(B^T B)^{-1} B^T$, respectively. We have

$$\begin{cases} NTPT^T v_A = \lambda^2 v_A \\ PT^T NT v_B = \lambda^2 v_B. \end{cases} \tag{7.3.3}$$

This reminds of the type of equation of a PCA with metrics (see Eq. (4.3.8)).

Therefore, let us denote, as in Chap. 4,

$$M = N^{1/2}, \qquad Q = P^{1/2}, \qquad R = MTQ \in \mathbb{R}^{p \times q}.$$

Then,

$$\begin{aligned} NTPT^{\mathrm{T}} &= M^2 T Q^2 T^{\mathrm{T}} \\ &= M(MTQ)(QT^{\mathrm{T}}M)M^{-1} \\ &= MRR^{\mathrm{T}}M^{-1}, \end{aligned}$$

and (the same for $PT^{\mathrm{T}}NT$) Eq. (7.3.3) reads

$$\begin{cases} MRR^{\mathrm{T}}M^{-1}v_{\mathrm{A}} = \lambda^2 v_{\mathrm{A}} \\ QR^{\mathrm{T}}RQ^{-1}v_{\mathrm{B}} = \lambda^2 v_{\mathrm{B}}. \end{cases} \tag{7.3.4}$$

Let us denote

$$w_{\mathrm{A}} = M^{-1}v_{\mathrm{A}}, \qquad w_{\mathrm{B}} = Q^{-1}v_{\mathrm{B}}.$$

Then, by left multiplication by M^{-1} of the first equation and by Q^{-1} of the second, we have

$$\begin{cases} RR^{\mathrm{T}}w_{\mathrm{A}} = \lambda^2 w_{\mathrm{A}} \\ R^{\mathrm{T}}Rw_{\mathrm{B}} = \lambda^2 w_{\mathrm{B}}, \end{cases} \tag{7.3.5}$$

where we recognize the PCA of $R = MTQ$. The complexity of this calculation is

$$\begin{array}{lll} T & = A^{\mathrm{T}}B & \text{is in} \quad \mathcal{O}(npq) \\ N & = (A^{\mathrm{T}}A)^{-1} & \mathcal{O}(np^2) \\ P & = (B^{\mathrm{T}}B)^{-1} & \mathcal{O}(nq^2) \\ M & = N^{1/2} & \mathcal{O}(p^3) \\ Q & = P^{1/2} & \mathcal{O}(q^3) \\ R & = MTQ & \mathcal{O}(p^2 q). \end{array}$$

Instead of computing the eigenvectors and eigenvalues of RR^{T} and $R^{\mathrm{T}}R$, we can compute the SVD of R. Let

$$R = W_{\mathrm{A}} \Lambda W_{\mathrm{B}}^{\mathrm{T}} \tag{7.3.6}$$

be this SVD. Then, $R^{\mathrm{T}}R = W_{\mathrm{B}} \Lambda^2 W_{\mathrm{B}}^{\mathrm{T}}$, so $R^{\mathrm{T}}RW_{\mathrm{B}} = W_{\mathrm{B}} \Lambda^2$, and the columns of W_{B} are the eigenvectors of $R^{\mathrm{T}}R$ associated with the eigenvalues in the diagonal of Λ^2. Similarly, $RR^{\mathrm{T}} = W_{\mathrm{A}} \Lambda^2 W_{\mathrm{A}}^{\mathrm{T}}$, so $RR^{\mathrm{T}}W_{\mathrm{A}} = W_{\mathrm{A}} \Lambda^2$, and the columns of W_{A} are the eigenvectors of RR^{T} associated with the eigenvalues in the diagonal of Λ^2.

We recognize in the PCA of $R = MTQ$ the first step of the PCA of T with metrics defined by $N = M^2$ on $\mathbb{R}^p$ and $P = Q^2$ on $\mathbb{R}^q$. Let us note also that w_A is a principal axis of the PCA of R^T, and w_B is a principal axis of the PCA of R. As $R \in \mathbb{R}^{p \times q}$ and as we have assumed that $p \geq q$, it is natural to run the PCA of R and hence compute the w_B as a principal axis of R. Then, w_A as a principal axis of R^T is a principal component of R and related to w_B by

$$w_A = Rw_B, \qquad \text{with} \quad \begin{cases} R & \in \mathbb{R}^{p \times q} \\ w_A & \in \mathbb{R}^p \\ w_B & \in \mathbb{R}^q. \end{cases} \tag{7.3.7}$$

Hence the SVD or search for eigenvalues and eigenvectors will be done once only.

- **Result:** The solution of the CCA of two arrays A and B is related to the solution of the PCAmet of $T = A^T B$ with an inner product defined by N on rows and by P on columns where $N = (A^T A)^{-1}$ and $P = (B^T B)^{-1}$. Let us define $M = N^{1/2}$, $Q = P^{1/2}$ and $R = MTQ$. Then, the solution of the PCAmet of (T, N, P) is given by the SVD of R as $R = W_A \Lambda W_B$. Then, if columns of W_A (respectively, W_B) are denoted w_A (respectively, w_B), we have

$$v_A = Mw_A, \qquad v_B = Qw_B, \tag{7.3.8}$$

and

$$y_A = Av_A, \qquad y_B = Bv_B. \tag{7.3.9}$$

It can be summarized as

A	$n \times p$	dataset
B	$n \times q$	dataset
T	$p \times q$	$A^T B$
M	$p \times p$	$(A^T A)^{-1/2}$
Q	$q \times q$	$(B^T B)^{-1/2}$
R	$p \times q$	MTQ
W_A, Λ^2, W_B	SVD	$R = W_A \Lambda^2 W_b^T$
V_A	$p \times p$	MW_A
V_B	$q \times q$	QW_B
Y_A	$n \times p$	AV_A
Y_B	$n \times q$	BV_B

These calculations can be presented in the following diagram:

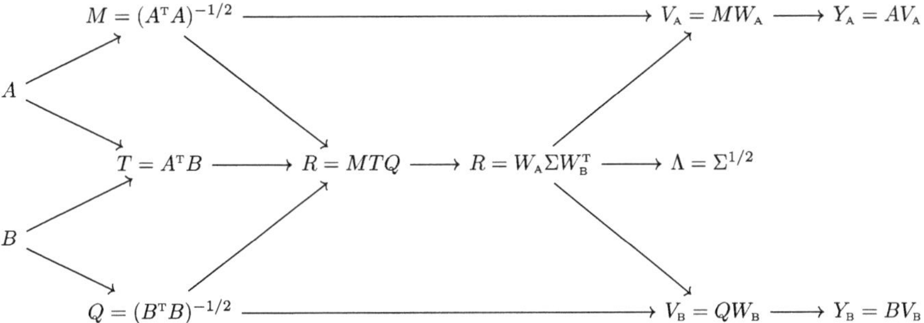

This result is far from anecdotal. It is the one which permits:

◇ To establish a deep connection between CoA and CCA
◇ To extend CCA to correlations between more than two arrays, as Multiple
Correlation Analysis

Both developments are presented in Chap. 8.

• **Algorithm:** There are several ways to write an algorithm for this calculation.
Here is a direct one, without calling PCA or PCA-MET.

Algorithm 10 Canonical Correlation Analysis: CCA(A, B)

1: **input** $A \in \mathbb{R}^{n \times p}$, $B \in \mathbb{R}^{n \times q}$ with $p \geq q$
2: **compute** $T = A^{\mathsf{T}} B$
3: **compute** $N = (A^{\mathsf{T}} A)^{-1}$
4: **compute** $P = (B^{\mathsf{T}} B)^{-1}$
5: **compute** $M = N^{1/2}$
6: **compute** $Q = P^{1/2}$
7: **compute** $R = MTQ$
8: **compute** $W_{\mathrm{A}}, \Sigma, W_{\mathrm{B}} = \mathrm{SVD}(R)$
9: **compute** $V_{\mathrm{A}} = MW_{\mathrm{A}}$
10: **compute** $V_{\mathrm{B}} = QW_{\mathrm{B}}$
11: **compute** $\Lambda = \Sigma^{1/2}$
12: **compute** $Y_{\mathrm{B}} = BV_{\mathrm{B}}$
13: **compute** $\Lambda = \Sigma^{1/2}$
14: **return** $Y_{\mathrm{A}}, Y_{\mathrm{B}}, V_{\mathrm{A}}, V_{\mathrm{B}}, \Lambda$

Comments Here are some comments on the algorithm:

→ The components y_{A} are the columns of Y_{A}.
→ The components y_{B} are the columns of Y_{B}.
→ The axis v_{A} are the columns of V_{A}.
→ The axis v_{B} are the columns of V_{B}.

$\rightarrow$ The singular values of R are the correlation coefficients λ, because $R^{\mathrm{T}} R w_{\mathrm{A}} = \lambda^2 w_{\mathrm{A}}$, and the eigenvalues of $R^{\mathrm{T}} R$ are the square of the singular values of R.

$\rightarrow$ It is possible to compute $M = N^{1/2}$ through an SVD of N: If $N = U_{\mathrm{M}} \Sigma_{\mathrm{M}} V_{\mathrm{M}}$, then $M = U_{\mathrm{M}} \Sigma_{\mathrm{M}}^{1/2} V_{\mathrm{M}}$ and Σ_{M} is diagonal. The same for computing Q from P.

Notes and References Canonical Correlation Analysis, or Canonical Analysis, seems to have been proposed by Hotelling in 1936 in Hotelling, H., Relation between two sets of variables, *Biometrika*, **28**:361–377. It is presented with a statistical approach in [And58, chap. 12] or [Rao73, Section 8f]. A review of Canonical Analysis with (debated) applications in ecology can be found in [Git85]. It is presented with a more algebraic approach in classical textbooks of the French school of data analysis, e.g., [LMT77, LMF82, EP90, Sap90] from which it has been borrowed and adapted here. Canonical Analysis is often referred to as Canonical Correlation Analysis (CCA). Both denominations will be used here.

Chapter 8
Multiple Canonical Correlation Analysis

Abstract This chapter looks at the simultaneous analysis of several (more than two) arrays of variables measured on the same set of individuals. One difficulty to overcome is that there is no canonical definition of a correlation between three or more variables. Also, there is no canonical extension to the correlation between three or more arrays. The commonly accepted idea is to construct a new array by concatenating column-wise the arrays to be analyzed simultaneously and perform a PCA of this large array. It can be shown that, in the case of two arrays, a given choice of metrics for the PCA makes it possible to recover the Canonical Analysis of two arrays. This choice naturally extends to more than two arrays and leads to the construction of Multiple Canonical Analysis.

Here, we develop a way to extend CCA to more than two arrays A and B. There are different ways to do it, which are not equivalent. Let us have three arrays, A, B, and C. Let us select one component per array, respectively, $y_A = Au_A$, $y_B = Bu_B$, and $y_C = Cu_c$ with $\|y_A\| = \|y_B\| = \|y_C\| = 1$. One can define three-ways CCA as finding a triplet (y_A, y_B, y_C) such that their correlation is maximal. There is however no canonical way to define the correlation between three vectors. It can be related to $\|y_A \wedge y_B \wedge y_C\|$ (as the norm of the wedge product is maximal when vectors are independent), or $\|y_A + y_B + y_C\|$, or other ways. Here, we extend CCA and CoA to more than two arrays thanks to an equivalence between CCA and an associated PCAmet (Sect. 8.2). We extend PCAmet to more than two arrays by columnwise concatenation of the arrays and the choice of a relevant inner product on the rows of the concatenated matrices (Sect. 8.3). Finally, an equivalence between CoA and CCA of two complete disjunctive arrays (Sect. 8.4) is shown. This leads by concatenation of more than two disjunctive arrays to MCoA, Multiple Correspondence Analysis (Sect. 8.5). . This is one way among others to link all these approaches, and a summary is given in Sect. 8.6.

8.1 Preamble

Let us recall the diagram of calculations of PCAmet presented in Sect. 4.3, page 48:

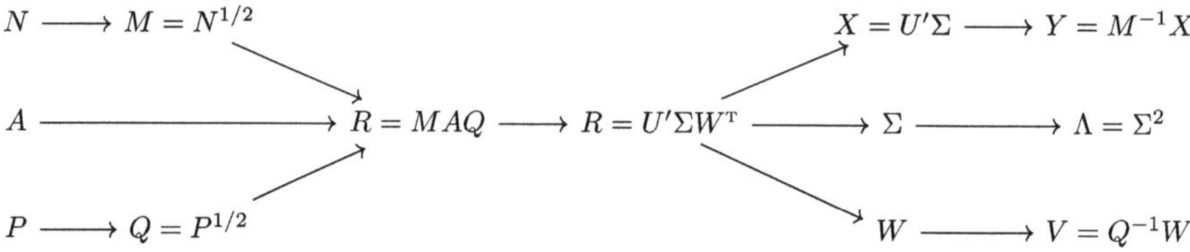

It resumes the calculation for PCA of A with inner products defined by N on rows and P on columns, denoted PCAmet(A, N, P).

Let us recall also the calculation for CCA of matrices A and B, presented in Sect. 7.3, page 87:

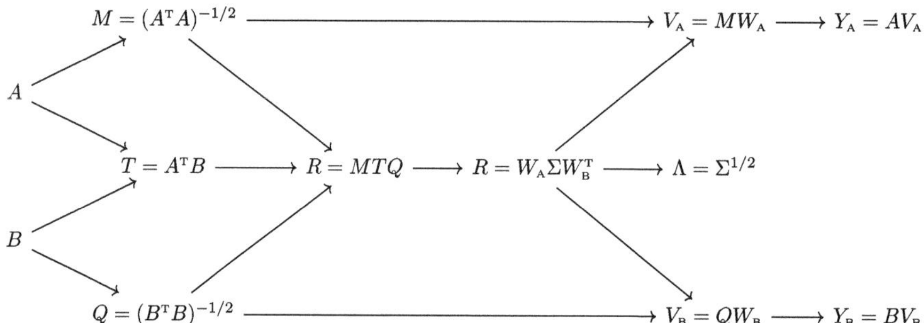

It resumes the calculation for CCA of the pair (A, B), denoted CCA(A, B).

We observe that running CCA(A, B) leads to performing the SVD of $R = MTQ$, and running PCAmet(A, N, P) leads to performing the SVD of $R = MAQ$. If we set $T = A^\mathsf{T}B, N = (A^\mathsf{T}A)^{-1}, P = (B^\mathsf{T}B)^{-1}$, both matrices R are equal: $R = MTA = MAQ$, and both approaches are equivalent. This section is based on this equivalence.

8.2 A Tight Link Between Canonical Analysis and PCA with Metric on Rows

Let $A \in \mathbb{R}^{n \times p}$ and $B \in \mathbb{R}^{n \times q}$ with $q \le p$ and $p + q \le n$ be two datasets, each set of quantitative variables (the columns of A and B) on the same set of items (the rows of A and B). Let us consider the Canonical Analysis of (A, B), which will be developed here with another calculation.

Therefore, let us consider the matrix

$$C = \begin{pmatrix} A & B \end{pmatrix} \qquad \in \mathbb{R}^{n \times (p+q)},$$

built by column-wise concatenation of A and B (denoted here and below blockwise as $(A \quad B)$). Let us define

$$\begin{vmatrix} N = (A^{\mathrm{T}} A)^{-1} & \in \mathbb{R}^{p \times p} \\ P = (B^{\mathrm{T}} B)^{-1} & \in \mathbb{R}^{q \times q}, \end{vmatrix}$$

and

$$D = \begin{pmatrix} N & 0 \\ 0 & P \end{pmatrix} \in \mathbb{R}^{(p+q) \times (p+q)}.$$

Let us perform the PCA of C with inner products on rows given by D (a row of C belongs to $\mathbb{R}^{p+q}$) and $\mathbb{I}_n$ on columns. Therefore, let us define $S = CD^{1/2}$. With a slight shift in notations when compared to those adopted in Chap. 4, i.e.,

$$M \to \mathbb{I}_n, \quad P \to D, \quad R \to S,$$

PCAmet of C with metrics on rows given by D, denoted PCAmet$(C, \mathbb{I}, D)$, leads to the PCA of $S = CD^{1/2}$, followed by some posttreatment. We have blockwise

$$S = CD^{1/2} = \begin{pmatrix} A & B \end{pmatrix} \begin{pmatrix} M & 0 \\ 0 & Q \end{pmatrix} = \begin{pmatrix} AM & BQ \end{pmatrix}, \tag{8.2.1}$$

with

$$S \in \mathbb{R}^{n \times (p+q)}, \qquad AM \in \mathbb{R}^{n \times p}, \qquad BQ \in \mathbb{R}^{n \times q}.$$

For running the PCA of S, it will be easier to use the eigen-decomposition of $S^{\mathrm{T}} S$, rather than the SVD decomposition of S. Principal axes v and principal components y are the solution of

$$S^{\mathrm{T}} S v = \lambda v, \qquad y = S v.$$

We have

$$S = \begin{pmatrix} AM & BQ \end{pmatrix}, \qquad S^{\mathrm{T}} = \begin{pmatrix} MA^{\mathrm{T}} \\ QB^{\mathrm{T}} \end{pmatrix}. \tag{8.2.2}$$

So,

$$S^{\mathrm{T}}S = \begin{pmatrix} MA^{\mathrm{T}}AM & MA^{\mathrm{T}}BQ \\ QB^{\mathrm{T}}AM & QB^{\mathrm{T}}BQ \end{pmatrix}. \tag{8.2.3}$$

Let us note that

$$MA^{\mathrm{T}}AM = \mathbb{I}_p, \qquad QB^{\mathrm{T}}BQ = \mathbb{I}_q, \tag{8.2.4}$$

because $M = (A^{\mathrm{T}}A)^{-1/2}$ (respectively, $Q = (B^{\mathrm{T}}B)^{-1/2}$). Let us denote

$$T = A^{\mathrm{T}}B.$$

Then

$$S^{\mathrm{T}}S = \begin{pmatrix} \mathbb{I}_p & MTQ \\ QT^{\mathrm{T}}M & \mathbb{I}_q \end{pmatrix}. \tag{8.2.5}$$

Let us decompose blockwise the axis v of the PCA as

$$v = \begin{pmatrix} v_{\mathrm{A}} \\ v_{\mathrm{B}} \end{pmatrix}.$$

Then, $S^{\mathrm{T}}Sv = \lambda v$ can be written

$$\begin{cases} v_{\mathrm{A}} & + MTQv_{\mathrm{B}} = \lambda v_{\mathrm{A}} \\ QT^{\mathrm{T}}Mv_{\mathrm{A}} + & v_{\mathrm{B}} = \lambda v_{\mathrm{B}}, \end{cases} \tag{8.2.6}$$

or

$$\begin{cases} MTQv_{\mathrm{B}} = (\lambda - 1)v_{\mathrm{A}} \\ QT^{\mathrm{T}}Mv_{\mathrm{A}} = (\lambda - 1)v_{\mathrm{B}}. \end{cases} \tag{8.2.7}$$

Let us denote

$$R = MTQ. \tag{8.2.8}$$

Then

$$\begin{cases} Rv_{\mathrm{B}} = (\lambda - 1)v_{\mathrm{A}} \\ R^{\mathrm{T}}v_{\mathrm{A}} = (\lambda - 1)v_{\mathrm{B}}, \end{cases} \tag{8.2.9}$$

or

$$\begin{cases} R R^{\mathrm{T}} v_{\mathrm{A}} = (\lambda - 1)^2 v_{\mathrm{A}} \\ R^{\mathrm{T}} R v_{\mathrm{B}} = (\lambda - 1)^2 v_{\mathrm{B}}. \end{cases} \tag{8.2.10}$$

The equation $R^{\mathrm{T}} R v_{\mathrm{B}} = (\lambda - 1)^2 v_{\mathrm{B}}$ is the PCA of R (this is the reason for the shift in notations when compared with Chap. 4 at the beginning of this section). We have $v_{\mathrm{B}} = R^{\mathrm{T}} v_a$ and $v_{\mathrm{A}} = R v_{\mathrm{B}}$. The principal components are solution of $y = S v$. If

$$y = \begin{pmatrix} y_{\mathrm{A}} \\ y_{\mathrm{B}} \end{pmatrix},$$

this yields

$$\begin{cases} y_{\mathrm{A}} = A M \, v_a \\ y_{\mathrm{B}} = B Q \, v_b. \end{cases} \tag{8.2.11}$$

We have $y_{\mathrm{A}} \in \mathrm{span}\, A$, $y_{\mathrm{B}} \in \mathrm{span}\, B$. As $y_{\mathrm{A}}, y_{\mathrm{B}}$ are principal components of a Canonical Analysis, we have $\|y_{\mathrm{A}}\| = \|y_{\mathrm{B}}\| = 1$, or

$$y_{\mathrm{A}} = \frac{A M \, v_a}{\|A M \, v_a\|}, \qquad y_{\mathrm{B}} = \frac{B Q \, v_b}{\|B Q \, v_b\|}. \tag{8.2.12}$$

One can check that they are the same component as the ones given for CCA, see Eq. (7.3.9), page 86, up to the normalization.

So, the CCA of (A, B) is equivalent (up to the transform $\lambda \mapsto (\lambda - 1)^2$) to the PCA of concatenated matrix $C = (A \quad B)$ with an inner product defined by D on rows. As concatenation can be extended to more than two arrays and that building D from N and P can be extended easily too, this leads naturally to MCCA presented in the next section.

8.3 Multiple Canonical Analysis

Knowing that, the extension of Canonical Analysis to more than two quantitative variables is straightforward.

Let us have m sets of variables labeled with $\ell \in [\![1, m]\!]$ defined on the same set of n items, each one given by a matrix A_ℓ with

$$A_\ell \in \mathbb{R}^{n \times p_\ell}, \qquad \sum_\ell p_\ell \leq n.$$

We denote $s = \sum_\ell p_\ell$. Let us define

$$M_\ell = (A_\ell^{\mathrm{T}} A_\ell)^{-1}, \quad \in \mathbb{R}^{p_\ell \times p_\ell}.$$

Let us build

$$C = \begin{pmatrix} A_1 & \dots & A_\ell & \dots & A_m \end{pmatrix},$$

(column-wise concatenation of matrices A_ℓ written blockwise) and

$$D = \begin{pmatrix} M_1 & 0 & \dots & \dots & 0 \\ 0 & M_2 & \ddots & \ddots & \vdots \\ \vdots & \ddots & \ddots & \ddots & \vdots \\ \vdots & & \ddots & \ddots & 0 \\ 0 & \dots & \dots & 0 & M_m \end{pmatrix}.$$

We have $C \in \mathbb{R}^{n \times q}$ and $D \in \mathbb{R}^{q \times q}$. Then, the MCCA of $C = (A_1 \ \dots \ A_m)$ is defined as the PCA of C with distances on rows given by D. It is the PCA of

$$S = \begin{pmatrix} A_1 M_1^{1/2} & \dots & A_m M_m^{1/2} \end{pmatrix}. \tag{8.3.1}$$

We have

$$S^{\mathrm{T}} = \begin{pmatrix} M_1^{1/2} A_1^{\mathrm{T}} \\ \vdots \\ M_m^{1/2} A_m^{\mathrm{T}} \end{pmatrix},$$

and $S^{\mathrm{T}} S$ can be written blockwise as

$$S^{\mathrm{T}} S = \begin{pmatrix} \mathbb{I}_{p_1} & R_{12} & \dots & R_{1m} \\ R_{21} & \mathbb{I}_{p_2} & R_{23} & R_{2m} \\ \vdots & \ddots & \ddots & \vdots \\ R_{m1} & \dots & \dots & \mathbb{I}_{p_m} \end{pmatrix}, \tag{8.3.2}$$

where

$$T_{\ell\ell'} = A_\ell^{\mathrm{T}} A_{\ell'} \quad \text{and} \quad R_{\ell\ell'} = M_\ell^{1/2} T_{\ell\ell'} M_\ell'^{1/2}.$$

So, we are looking for the eigen-decomposition of $S^\mathsf{T}S$, i.e., finding eigenpairs (λ, v) with $\lambda \in \mathbb{R}$ and $v \in \mathbb{R}^s$ (we recall that $s = \sum_\ell p_\ell$) such that

$$S^\mathsf{T}Sv = \lambda v. \tag{8.3.3}$$

v can be written blockwise as

$$v = \begin{pmatrix} v_1 \\ \vdots \\ v_\ell \\ \vdots \\ v_m \end{pmatrix},$$

where v_ℓ is the canonical axis for array A_ℓ. In classical PCA, y, the principal component, is given by $y = Sv$. Here, in MCCA, we have one component y_ℓ per array A_ℓ, of norm one, given by

$$y_\ell = \frac{A_\ell M_\ell^{1/2} v_\ell}{\|A_\ell M_\ell^{1/2} v_\ell\|}. \tag{8.3.4}$$

Let us write it for $m = 3$, to see the difficulty to overcome. We have

$$v = \begin{pmatrix} v_1 \\ v_2 \\ v_3 \end{pmatrix},$$

and

$$\begin{cases} & R_{12}v_2 + R_{13}v_3 = (\lambda - 1)v_1 \\ R_{21}v_1 + & R_{23}v_3 = (\lambda - 1)v_2 \\ R_{31}v_1 + R_{32}v_2 & = (\lambda - 1)v_3. \end{cases} \tag{8.3.5}$$

It is not easy to find an analytical solution for each axis v_ℓ knowing the matrices $R_{\ell\ell'}$. So, Eq. (8.3.3) has to be solved numerically: It yields vector $v \in \mathbb{R}^q$, and axis $v_\ell \in \mathbb{R}^{p_\ell}$ is obtained by blockwise decomposition of v. This yields all axes v_ℓ.

It seems natural to solve $S^\mathsf{T}Sy = \lambda y$ rather than $SS^\mathsf{T}v = \lambda v$, because $S \in \mathbb{R}^{n \times s}$ with $s < n$. But solving $SS^\mathsf{T}y = \lambda y$ will shed some light on the link between the

PCA of C with row-wise metrics given by D and CCA of the arrays A_ℓ. Indeed, we have

$$
\begin{aligned}
SS^{\mathrm{T}} &= \sum_{\ell=1}^{m} A_\ell M_\ell^{1/2} M_\ell^{1/2} A_\ell^{\mathrm{T}} \\
&= \sum_{\ell=1}^{m} A_\ell (A_\ell^{\mathrm{T}} A_\ell)^{-1} A_\ell^{\mathrm{T}} \\
&= \sum_{\ell=1}^{m} \mathcal{P}_\ell,
\end{aligned}
\tag{8.3.6}
$$

where $\mathcal{P}_\ell$ is the projection on span A_ℓ in $\mathbb{R}^n$. We then have

$$
\left(\sum_{\ell=1}^{m} \mathcal{P}_\ell \right) y = \lambda y.
\tag{8.3.7}
$$

As $\mathcal{P} = \sum_\ell \mathcal{P}_\ell$ is symmetric $((SS^{\mathrm{T}})^{\mathrm{T}} = SS^{\mathrm{T}})$, the largest eigenvalue of $\mathcal{P}$ verifies

$$
\lambda = \max_{\|y\|=1} \|\mathcal{P}y\|.
\tag{8.3.8}
$$

So, the first component y is the vector in $\mathbb{R}^n$ of norm one such that $\|\sum_\ell \mathcal{P}_\ell y\|$ is maximal. It is the solution of the CCA of $(A_1 \ \ldots \ A_m)$.

8.4   A Tight Link Between Canonical Analysis and Correspondence Analysis

Let us consider a set of two qualitative variables A and B on the same set of items $[\![1, n]\!]$. We build the so-called indicator array of each variable, called as well complete disjunctive table, or one hot encoding more recently (in machine learning). If there are n items, p modalities for A, and q for B, it is an $n \times p$ array for A (respectively, $n \times q$ for B), with, in each row i, zero in all columns but one in column j if the modality j (respectively, k) of the variable has been observed for item i in A (respectively, in B). If $A = (a_{ij})_{i,j}$ and $B = (b_{ik})_{i,k}$, we have

$$
a_{ij} = \begin{cases} 1 & \text{if modality } j \text{ has been observed for item } i \\ 0 & \text{otherwise,} \end{cases}
$$

$$
b_{ik} = \begin{cases} 1 & \text{if modality } k \text{ has been observed for item } i \\ 0 & \text{otherwise.} \end{cases}
$$

Let us perform the Canonical Analysis of the pair of arrays (A, B). Therefore, following the diagram in Sect. 8.1, we define

$$T = A^{\mathrm{T}}B, \qquad D_{\mathrm{A}} = (A^{\mathrm{T}}A)^{-1}, \qquad D_{\mathrm{B}} = (B^{\mathrm{T}}B)^{-1}, \qquad R = D_{\mathrm{A}}^{1/2} T D_{\mathrm{B}}^{1/2},$$

where D_{A} plays the role of N and D_{B} of P (the reason for this shift in notations will be clear below). The solution of CCA is obtained by doing the SVD of R (see Chap. 7) with $R = W_{\mathrm{A}} \Lambda W_{\mathrm{B}}$.

Key Observations Let us make three observations:

→ $(A^{\mathrm{T}}A)^{-1} \in \mathbb{R}^{p \times p}$ (respectively, $(B^{\mathrm{T}}B)^{-1} \in \mathbb{R}^{q \times q}$) is the diagonal matrix with in position (j, j) (respectively, (k, k)) the inverse of the number of rows in A (respectively, B), where the modality j (respectively, k) of the variable has been observed (hence the notations D_{A} and D_{B}).
→ One recognizes in T the contingency table of A and B.
→ Then, D_{A} (respectively, B_{B}) is the diagonal matrix of the inverse of the row margins (respectively, of the column margins) of T.

- **Result:** Performing the CCA of (A, B) is performing the PCA of $R = D_{\mathrm{A}}^{1/2} T D_{\mathrm{B}}^{1/2}$. One can recognize in the PCA of R the PCA of T with inner product on columns defined by D_{A} and on rows by D_{B}. This is the CoA of T (up to the normalization to frequencies, see Sect. 5.3). This equivalence between the CCA of (A, B) and the CoA of $T = A^{\mathrm{T}}B$ is the main result through which CoA can be extended to more than two variables, as MCoA.

This result is natural and can be understood as follows. We know from Sect. 8.2 that the CCA of (A, B) is equivalent to the PCAmet of $(A \quad B)$ with an inner product on rows given by

$$D = \begin{pmatrix} D_{\mathrm{A}} & 0 \\ 0 & D_{\mathrm{B}} \end{pmatrix}.$$

Let $C = (A \quad B)$. Let j be any column of C. D is the diagonal matrix with, in position (j, j), the inverse of the margin of column j:

$$D[j, j] = \frac{1}{c_j}, \qquad c_j = \sum_{i=1}^{n} C[i, j]. \tag{8.4.1}$$

It is the matrix D_c in CoA. Similarly, the row margins $r_i = \sum_j [C[i, j]$ are all equal: For any i, $r_i = 2$. As all weights on rows are equal, they can be set to one. So we can define $D_r = \mathbb{I}_n$.

- **Result:** The CCA of (A, B) is the PCAmet of array $(A \quad B)$ with inner product defined by D_c on rows and $\mathbb{I}_n = D_r$ on columns. It is the CoA of $(A \quad B)$ even if $(A \quad B)$ is not a contingency table.

Then, a pair (A, B) of two indicator arrays on two variables on the same set of items being given, this shows the equivalence between:

$\rightarrow$ The CCA of (A, B)
$\rightarrow$ The CoA of $T = A^{\mathrm{T}}B$
$\rightarrow$ The CoA of $(A \quad B)$

MCoA is based on these equivalences and MCCA.

8.5   Multiple Correspondence Analysis

This equivalence between Correspondence Analysis and Canonical Analysis leads to an extension of Correspondence Analysis to more than two variables. Let us note however that the posttreatment in CoA and this CCA are different (see the diagrams in the preamble, Sect. 8.1). Therefore, we will use the equivalence between the CCA and the PCAmet which extends CCA to more than two arrays.

Let us have m categorical variables observed each on n items. Let us denote by A_ℓ with $\ell \in [\![1, m]\!]$ the indicator array of variable ℓ, i.e., is an $n \times p_\ell$ binary array with

$$a_{ij_\ell} = \begin{cases} 1 & \text{if modality } j_\ell \text{ has been observed for item } i \\ 0 & \text{otherwise.} \end{cases}$$

Let us define

$$C = \left(A_1 \ldots A_\ell \ldots A_m\right),$$

and the metric on rows of C defined by the matrix D blockwise diagonal is defined as

$$D = \mathrm{Diag}(D_\ell) \qquad \text{with} \quad D_\ell = (A_\ell^{\mathrm{T}} A_\ell)^{-1},$$

or

$$D = \begin{pmatrix} D_1 & 0 & \ldots & \ldots & 0 \\ 0 & \ddots & \ddots & & \vdots \\ \vdots & \ddots & D_\ell & \ddots & \vdots \\ \vdots & & \ddots & \ddots & 0 \\ 0 & \ldots & \ldots & 0 & D_m \end{pmatrix}. \tag{8.5.1}$$

Then, the Multiple Correspondence Analysis of $(A_1, \ldots, A_m)$ is equivalent to the PCAmet of C with metric defined by D on rows.

Each matrix D_ℓ is diagonal. Let us consider the variable ℓ coded by array A_ℓ. Then, D_ℓ is a diagonal matrix of weights, with weight $w^\ell = (w_1^\ell, \ldots, w_{p_\ell}^\ell)$, where w_k^ℓ is the number of items in array A_ℓ which have the modality k for the variable. Then, matrix D itself is a diagonal matrix. So MCoA can be reduced to a PCA with weights on columns.

8.6 Summary of Relationships Between Some Methods

We have shown that:

1. Given two arrays A, B, the CCA of (A, B) is equivalent to the PCAmet of $(A \quad B)$ with metrics defined by D on rows, where D is the block diagonal with $(A^{\mathrm{T}}A)^{-1}$ and $(B^{\mathrm{T}}B)^{-1}$ (Sect. 8.2).
2. This leads to the extension of CCA to MCCA by column-wise concatenation of all arrays and performing a PCAmet with metrics on rows defined by an extended block diagonal matrix (Sect. 8.3).
3. If A, B are two complete disjunctive arrays each of one variable on the same set of items, then the CCA of (A, B) is equivalent to the PCAmet of $T = A^{\mathrm{T}}B$ with metrics on columns defined by $(A^{\mathrm{T}}A)^{-1}$ (which is diagonal) and on columns by $(B^{\mathrm{T}}B)^{-1}$, where we recognize the CoA of T (Sect. 8.4).
4. This leads to MCoA as Correspondence Analysis of several arrays, with the PCAmet of their column-wise concatenation as MCCA of the set of arrays (Sect. 8.5).

Notes and References These links between different treatments of the same dataset which establish some dependencies between methods have been subject to thorough studies and presentations by the French school of multivariate analysis, more algebraic and geometrical than statistical, in the 1970s, with the names of B. Escofier, J.-P. Fénelon, L. Lebart, A. Morineau, and J. Pagès, among others. It is presented in all textbooks of this school in multivariate data analysis, under chapters called "other methods and complements," like [LMT77, LMF82, LMP00]. Since then, the diversity of related methods has flourished, and a more recent panorama is given in [GB06]. According to the introduction of [GB06], L. Guttman should be credited for all the basic ideas at the root of Multiple Correspondence Analysis, in Guttman, L. (1941) *The quantification of a class of attributes: A theory and method of scale construction*, Chapter in *The Prediction of Personal Adjustment*, P. Horst, eds. New York: Social Science Research Council. The article [TY85] is a thorough survey of different methods associated with Multiple Correspondence Analysis, with both a historical background (and they point out several independent beginnings in the 1930s and early 1940s) and, long before the present notes, an organization of methods to show that they all lead to the same equation to analyze the data. The unifying formalism selected in [TY85] is the duality diagram.

Chapter 9
Multidimensional Scaling

Abstract This chapter deals with a slightly different approach, but still associated with PCA. The dataset is not a set of variables and features, or a contingency table, but a pairwise distance array between some items. It is assumed implicitly that the distance array contains some information about the structure of the set of items. To reveal this structure, the objective is to recover a point cloud in a Euclidean space with a one-to-one correspondence between items and points, such that the Euclidean distance between points mimics as much as possible the distances as given by the distance array. Then, it is possible to derive a low-dimensional approximation of the point cloud with PCA, which offers the possibility to study its structure in a low-dimensional space, i.e., in a much easier way than in a high-dimensional space. Not all distance arrays can lead to such a procedure: In some cases, there exists no point cloud in a Euclidean space which can exactly mimic the distances as given. It is shown that, in such a case, a quadratic embedding provides a way to associate with the distance array a point cloud in a quadratic space endowed with a pseudo-Euclidean structure. How to reveal a shape in a pseudo-Euclidean space seems to be an open question, much less mature than in a Euclidean space.

Multidimensional Scaling is a technique to map a discrete metric space into a Euclidean space. Let (M, d) be a metric space with $|M| = n$. It is given by a pairwise distance matrix

$$D = (d_{ij})_{i,j} \in \mathbb{R}^{n \times n} \qquad \text{with} \quad 1 \le i, j \le n,$$

such that

$$d_{ij} = d(i, j).$$

MDS at dimension r addresses the question of finding a point cloud

$$X = (x_i)_{1 \le i \le n}, \qquad x_i \in \mathbb{R}^r,$$

such that the distance between points x_i and x_j is as close as possible to d_{ij} or, loosely speaking,

$$\|x_i - x_j\| \approx d_{ij}.$$

A matrix $X \in \mathbb{R}^{n \times r}$ is attached to the point cloud $\mathcal{X}$, with x_i being row i of X.
   Two situations may occur:

1. Either the distances d_{ij} come from a Euclidean distance between (unknown) points, and the problem is to recover them, i.e., produce an isometry, and a best approximation of it at dimension r
2. Or they do not come from a Euclidean distance, and a best approximation is sought for, knowing there exists no exact isometry.

   Problem 1 is known as *classical MDS* and problem 2 as *Least Square Scaling*. In this note, we address the problem of classical MDS only. Least square scaling is a delicate and nontrivial optimization problem.
   It appears however that, in practical cases, i.e., when given with a distance matrix D, one never knows whether the distances come from a Euclidean point cloud or not, and classical MDS assumes implicitly that the distances in metric space (M, d) are ℓ^2-norm distance in a (unknown) Euclidean space. However, it is not always the case. Here is a counterexample. Let us have a graph $G = (V, E)$ as a star with four vertices $V = \{c, i, j, k\}$ and three edges $E = \{(c, i), (c, j), (c, k)\}$:

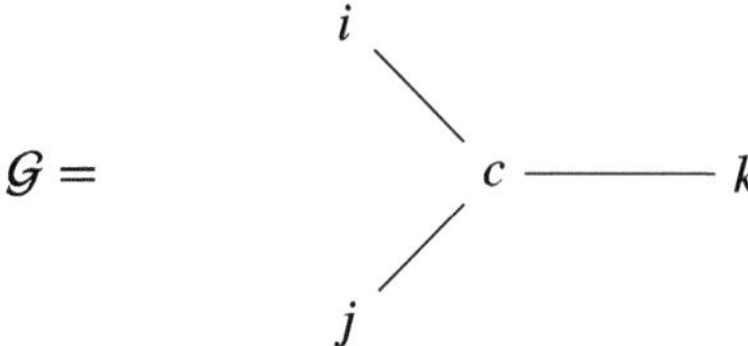

Distances between vertices are taken as the length of the shortest path on the graph, with equal weight $w = 1$ on each edge. They are given by the matrix D as follows:

	c	i	j	k
c	0	1	1	1
i	1	0	2	2
j	1	2	0	2
k	1	2	2	0

   One can show by elementary construction that there exist no dimension $n \in \mathbb{N} \cup \infty$ and Euclidean (Hilbert) space of dimension n with four points $(i, j, k, c) \in \mathbb{R}^n$ such that their pairwise Euclidean distances are given by matrix D. Let $n = 2$, and select $c = (0, 0)$ and $k = (1, 0)$ without loss of generality. Then, i and j are on the circle of center $c = (0, 0)$ and radius 1, because $d(i, c) = d(j, c) = 1$, and on the circle of center $k = (1, 0)$ and radius 2, because $d(i, k) = d(j, k) = 2$. These two circles intersect at one point (where they are tangent) $(-1, 0)$. Then, $i = j = (-1, 0)$ and $d(i, j) = 0$. The same demonstration holds for $n = 3$ with intersection of spheres.

Notes and References There exist many excellent textbooks presenting MDS, classical MDS, or LSS. We can recommend [CC01] or [Ize08, chap. 13]. A comprehensive reference is [BG05]. A classical and rigorous reference with many results, their demonstration, and history is [MKB79] which we highly recommend for those enjoying a mathematical-based approach. Classical MDS has been proposed by Torgerson in 1952 [Tor52]. Here, we have followed [CC01, chap. 2].

9.1 Overview

Provided the distances are Euclidean, three points with pairwise distances can be arranged as a triangle in $\mathbb{R}^2$, and four points with pairwise distances can be arranged as a tetrahedron in $\mathbb{R}^3$. More generally, n points with pairwise Euclidean distances can be isometrically embedded in $\mathbb{R}^{n-1}$ at most. Then, MDS at rank r is made in two steps:

1. Find a point cloud X in $\mathbb{R}^n$ with distances matching exactly the values of $d(i, j)$ for each pair.
2. Reduce the dimension $r < n$ of the space where X lives, and build the point cloud X_r as close as possible to X. This can be solved by PCA of X, the matrix attached to X. It will appear that principal axes and components of the PCA of X can be given by MDS without further calculations.

Let $X = (x_i)_i$ be a cloud on n points in $\mathbb{R}^r$ with distances

$$\|x_i - x_j\| = d_{ij}.$$

The Gram matrix $G \in \mathbb{R}^{n \times n}$ of X is the matrix with elements

$$G_{ij} = \langle x_i, x_j \rangle,$$

or

$$G = XX^\mathsf{T}.$$

A Gram matrix is symmetric, semi-definite, and positive. It is definite if and only if the vectors $(x_i)_i$ are independent. Conversely, every matrix that is symmetric, semi-definite, and positive is the Gram matrix of a given, family of vectors $(x_i)_i$. Classical MDS is precisely about finding such a family knowing the Gram matrix. The solution is not unique because, if X is a solution, any family X' obtained from

X by an isometry on the rows, which leaves the inner product invariant, is a solution too.

There is a well-known correspondence between the Gram Matrices G of inner products $\langle x_i, x_j \rangle$ and the Euclidean Distance Matrix of quantities $\|x_i - x_j\|$, both $n \times n$. If

$$\left|\begin{aligned} g_{ij} &= \langle x_i, x_j \rangle \\ d_{ij}^2 &= \|x_i - x_j\|^2, \end{aligned}\right.$$

then

$$\begin{cases} d_{ij}^2 = g_{ii} + g_{jj} - 2g_{ij} \\ g_{ij} = -\dfrac{1}{2}\left(d_{ij}^2 - d_{i\bullet}^2 - d_{j*}^2 + d_{\bullet\bullet}^2\right), \end{cases} \tag{9.1.1}$$

with

$$\begin{cases} d_{i\bullet}^2 = \dfrac{1}{n}\sum_j d_{ij}^2 \\ d_{\bullet\bullet}^2 = \dfrac{1}{n^2}\sum_{i,j} d_{ij}^2 = \dfrac{1}{n}\sum_i d_{i\bullet}^2. \end{cases} \tag{9.1.2}$$

Such a correspondence has been studied for decades (see, e.g., [Sch38, Lau98]). We then have the scheme (with an arrow meaning "built from"):

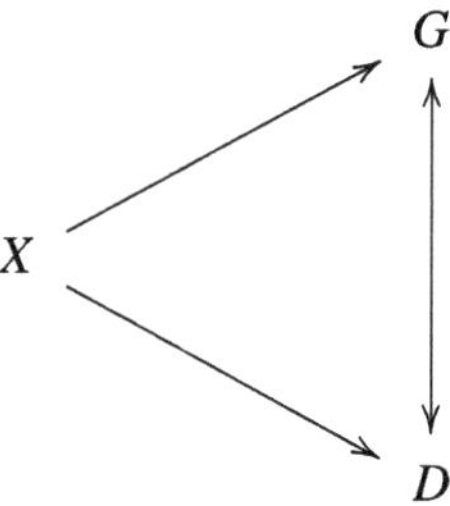

A key question is to know whether one-way arrows $X \longrightarrow G$ and $X \longrightarrow D$ can be reversed, i.e., whether X can be computed knowing G or knowing D. This amounts to answering to the question: A pairwise distance matrix D being given, is there a dimension m and a point cloud X in $\mathbb{R}^m$ such that the distance between x_i and x_j is precisely d_{ij}?

A matrix D being given, it is always possible to compute a matrix G by Eq. (9.1.1). But G is not necessarily positive, i.e., it is not necessarily the Gram matrix of a point cloud X. The conditions on D for G to be positive, i.e., a Gram matrix, have been thoroughly studied. In most of the cases, when the Gram matrix is not positive, the negative eigenvalues are just ignored. This leads to some subtleties when connecting the EVD and the SVD of the Gram matrix to compute the coordinates.

The coordinates of the point cloud X can be computed from the eigenvectors and eigenvalues or the Singular Value Decomposition of the Gram matrix. The recipe is given here.

9.2 Setting the Problem

Let D be a discrete metric space, with $|M| = n$. We assume that there exists (unknown) point clouds $\mathcal{X}$ of n points in $\mathbb{R}^n$ such that, if x_i is the point i,

$$\forall\, i, j, \quad \|x_i - x_j\| = d_{ij}.$$

We denote by $X \in \mathbb{R}^{n \times n}$ the matrix associated with $\mathcal{X}$. Let us denote by G the Gram matrix of X, i.e., $g_{ij} = \langle x_i, x_j \rangle$, or

$$G = XX^{\mathrm{T}}. \tag{9.2.1}$$

The objective of MDS is to compute X knowing D. Of course, the solution is not unique: If X is a solution, each matrix X' where x'_i is built from x_i by the action of a rotation is a solution too. So, the objective is to compute one solution. A further step is to constrain the rank of X, or to bound the dimension of the space in which $\mathcal{X}$ lives, and find the best approximation X_r of rank r of X (this in PCA). It amounts to find a point cloud in a low-dimensional space with distances between points being as close as possible to the distances between elements in metric space. This can be set as

Given	(M, d) a discrete metric space
	with $|M| = n$,
	d coming from an unknown Euclidean distance,
	r with $1 \le r \le n$
Set	$D \in \mathbb{R}^{n \times n}$ as the distance matrix between points in (M, d)
Compute	$G \in \mathbb{R}^{n \times n}$ as the Gram matrix of D as in (9.1.1)
Find	$X \in \mathbb{R}^{n \times n}$
such that	$G = XX^{\mathrm{T}}$
and	X_r as best approximation of X at rank r.

9.3 Solving the Problem

Then, the problem can be solved either with eigen-decomposition of G or with its SVD. Let us present both approaches together.

Let $(u_\alpha, \lambda_\alpha)_\alpha$ be the set of eigenpairs of G

$$Gu_\alpha = \lambda_\alpha u_\alpha, \tag{9.3.1}$$

with

$$\lambda_1 \geq \lambda_2 \geq \ldots \geq \lambda_n \geq 0.$$

As G is symmetric, the eigenvectors if normed form an orthonormal family. If $U = [u_1 | \ldots | u_n]$ is the matrix with u_α in column α and Λ the diagonal matrix with $(\lambda_\alpha)_\alpha$ in its diagonal, we have

$$GU = U\Lambda. \tag{9.3.2}$$

As U is orthonormal, $UU^\mathsf{T} = \mathbb{I}_n$, and we have by right multiplication by U^T

$$G = U\Lambda U^\mathsf{T}. \tag{9.3.3}$$

We recognize here the SVD of G if G is positive. The case where G is not positive is handled by designing a quadratic embedding and is addressed in Sect. 9.5. Let us note that the standard practice when G is not positive is not to design a quadratic embedding, but to clip to zero the nonpositive eigenvalues and eigenvectors, i.e., to keep track of positive eigenvalues only and associated eigenvectors. If G is positive, the eigenvalues of G are its singular values, and if G is not positive, the negative eigenvalues are singular values up to their sign (if $\lambda < 0$ is an eigenvalue of G, $-\lambda > 0$ is a singular value of G).

As G is definite positive, let

$$\Sigma = \Lambda^{1/2}.$$

Then

$$G = U\Lambda U^\mathsf{T} = U\Sigma^2 U^\mathsf{T} = (U\Sigma)(U\Sigma)^\mathsf{T}, \tag{9.3.4}$$

and we can select

$$X = U\Sigma, \qquad \Sigma = \Lambda^{1/2}. \tag{9.3.5}$$

It is not the only solution, because any matrix $X' = X\Omega$, where Ω is a rotation in $\mathbb{R}^m$, is a solution too.

9.4   Dimension Reduction

Once matrix X has been computed, finding a point cloud X_r of n points in $\mathbb{R}^r$ as close as possible to $\mathcal{X}$ is done by the PCA of X.

Let us recall that

$$X = U\Sigma \qquad \text{with} \quad U^{\mathsf{T}}U = \mathbb{I}_n. \tag{9.4.1}$$

We recognize here the SVD of X as $X = U\Sigma V^{\mathsf{T}}$, with singular values being the values in the diagonal of Σ and $V = \mathbb{I}_n$. Hence, $X = U\Sigma$ is the matrix of principal components of X, and the dimensionality reduction of X by PCA is given as a cherry on top. Indeed, $X = U\Sigma$ can be written

$$X = \sum_{\alpha=1}^{n} \sigma_\alpha\, u_\alpha \otimes e_\alpha \tag{9.4.2}$$

if $(e_\alpha)_\alpha$ is the standard orthonormal basis of $\mathbb{R}^n$ (e_α is column α of V), with

$$\sigma_1 \geq \sigma_2 \geq \ldots \geq \sigma_n \geq 0,$$

and

$$X_r = \sum_{\alpha=1}^{r} \sigma_\alpha u_\alpha \otimes e_\alpha. \tag{9.4.3}$$

Algorithm 11 Classical MDS: $X, \Sigma = \mathrm{D}, r$

1: **input:** a distance matrix $D \in \mathbb{R}^{n \times n}$; a dimension $r < n$
2: **compute** the Gram matrix of D: $G = \mathrm{GRAM}(D)$
3: **compute** the eigen-pairs $(u_\alpha, \lambda_\alpha)$ such that $Gu_\alpha = \lambda_\alpha u_\alpha$
4: keep in U the columns associated with nonnegative eigenvalues of G; clip them off in Λ
5: **compute** $\Sigma = \Lambda^{1/2}$
6: **compute** $X = U\Sigma$
7: keep in X_r the r first columns of X only, and in Λ_r the first r nonnegative eigenvalues of G
8: **return** X_r, Λ_r

Wrapping all this together yields the algorithm 11 for MDS:

- **Note:** If the user selects the computation of the eigenvalues of G, it is simple to detect those which are negative and clip the corresponding columns in U (and values in Λ). If the SVD is selected, we have $G = U\Sigma V^{\mathsf{T}}$ (classical notation $(U, \Sigma, V$, here $\Sigma \neq \Lambda^{1/2})$, with columns of V (respectively, singular values) being the same as the columns of U (respectively, eigenvalues of G) up to the sign, depending on the sign of the corresponding eigenvalue of G:

$$\lambda_\alpha > 0 \implies v_\alpha = u_\alpha \quad \sigma_\alpha = \lambda_\alpha$$
$$\lambda_\alpha < 0 \implies v_\alpha = -u_\alpha \quad \sigma_\alpha = -\lambda_\alpha.$$

Summary Here is a summary of the calculations:
For Gram Matrix $(G = \text{GRAM}(D))$

D	$n \times n$	distance matrix	
$d_{i\bullet}^2$	$\forall\, i$		$d_{i\bullet}^2 = \frac{1}{n} \sum_j d_{ij}^2$
$d_{\bullet j}^2$	$\forall\, j$		$d_{\bullet j}^2 = \frac{1}{n} \sum_i d_{ij}^2$
$d_{\bullet\bullet}^2$			$d_{\bullet\bullet}^2 = \frac{1}{n^2} \sum_{i,j} d_{ij}^2$
G	$n \times n$	Gram matrix	$g_{ij} = -\frac{1}{2}(d_{ij}^2 - d_{i\bullet}^2 - d_{\bullet j}^2 + d_{\bullet\bullet}^2)$

Then

D	$n \times n$	distance matrix	
G	$n \times n$	Gram or kernel matrix of D	$G = \text{GRAM}(D)$
U, Λ		SVD of G	$G = U \Lambda U^\mathrm{T}$
Σ			$\Sigma = \Lambda^{1/2}$
X	$n \times n$	point clouds	$X = U\Sigma$

SVD with Randomized SVD In Sect. 2.4, we have seen how an SVD can be computed for large dimensions with randomized SVD. We present here how it can be adapted for the SVD of the Gram matrix. Therefore, we can choose a random matrix Ω, compute $Y = G\Omega$, Q such that $Y = QR$, define $G' = QQ^\mathrm{T}G \approx G$, and do the SVD of G'. If $B = Q^\mathrm{T}G$, the SVD of B is $B = U_\mathrm{B}\Sigma V^\mathrm{T}$ and $G' = QB = U\Sigma V^\mathrm{T}$ with $U = QU_\mathrm{B}$. However, one cannot write $G' = X'X'^\mathrm{T}$ because G' is not symmetric, and $V \neq U$. To solve this, one can do a double projection of G, row-wise and column-wise, and define $G'' = (QQ^\mathrm{T})G(QQ^\mathrm{T})$, with still $G'' \approx G$. One defines $C = Q^\mathrm{T}GQ$, which is symmetric, and does its SVD by $C = U_\mathrm{C}\Sigma U_\mathrm{C}^\mathrm{T}$; hence $G'' = U\Sigma U^\mathrm{T}$ with $U = QU_\mathrm{C}$. One can then write $G \approx X''X''^\mathrm{T}$ with $X'' = U\Sigma^{1/2}$. This can be derived as follows:

Algorithm 12 SVD of a Gram matrix with Gaussian Random Projection $\text{SVD\_GRP\_GRAM}(G, k))$

1: **input** $G \in \mathbb{R}^{n \times n}$, k as prescribed rank
2: **build** $\Omega \in \mathbb{R}^{n \times k}$, random $(\Omega[i, j] \sim \mathcal{N}(0, 1))$
3: **compute** $Y = G\Omega$
4: **compute** the $QR-$decomposition of Y: $Y = QR$
5: **build** $C = Q^{\mathsf{T}}GQ$
6: **run** the SVD of C: $C = U_{\mathrm{C}}\Sigma U_{\mathrm{C}}^{\mathsf{T}}$,   or $(U_{\mathrm{C}}, \Sigma, U_{\mathrm{C}}) = \text{SVD}(C)$
7: **compute** $U = QU_{\mathrm{C}}$
8: **return** U, Σ

Here are the dimensions of the involved matrices:

Matrix	Dimensions	Computation
G	$n \times n$	Gram matrix
Ω	$n \times k$	Gaussian random matrix
Y	$n \times k$	$Y = G\Omega$
Q	$n \times k$	$Y = QR$
C	$k \times k$	$C = Q^{\mathsf{T}}GQ$
U_{C}	$k \times k$	$C = U_{\mathrm{C}}\Sigma U_{\mathrm{C}}^{\mathsf{T}}$
Σ	$k \times k$	idem
U	$n \times k$	$U = QU_{\mathrm{C}}$

One observes that the complexity of the calculation of the SVD of C is in $O(k^3)$, whereas the calculation of $Y = G\Omega$ is in $O(n^2 k)$ and more "expensive."

9.5 Quadratic Embedding

There is still one point to look at. In all these developments, we have assumed that the Gram matrix G built from distance matrix D with recipe in Eq. (9.1.1) is positive, i.e., that all eigenvalues of G (the λ_α) are nonnegative. This is a property of a Gram matrix but is not always the case for the matrix G computed with real data. If one eigenvalue at least is negative, the matrix G is more strictly called a *kernel matrix*, and there is no isometry between (M, d) and a Euclidean space, whatever its dimension. However, it can be shown that there exists a quadratic embedding between (M, d) and a pseudo-Euclidean space, i.e., a vector space with a quadratic form with a signature (p, m) (see Appendix B.3 for a short introduction to quadratic forms and spaces). Such an embedding is rarely done, and most of the time the axes (and components) associated with negative eigenvalues of G are simply ignored or clipped to zero. We show in this section how to take into account the axes associated with negative eigenvalues of the kernel matrix.

Quadratic Embedding Let (M, d) be a discrete metric space, with $M =$
$\{1, \ldots, i, \ldots, n\}$. Let (E, q) be a quadratic space, with $\dim E = n$, and φ a
map

$$M \xrightarrow{\;\varphi\;} E.$$

Let us denote

$$a_i = \varphi(i).$$

So, $a_i \in E$, and we have a cloud of n points in E, each one corresponding to an
element in M. φ is a quadratic embedding or preserves the distances, if

$$q(a_i - a_j) = d^2(i, j). \tag{9.5.1}$$

If (E, q) is a Euclidean space, $q(a_i - a_j) = \|a_i - a_j\|^2$, and we recover $\|a_i - a_j\|^2 = d_{ij}^2$. We show next:

⋄ That such an embedding exists (and is not unique)
⋄ How to build it by specifying q and φ

We first build q and next φ.

Building a Quadratic Form The key is that the kernel matrix G built with recipe
in Eq. (9.1.1) is the polar form of q, which permits to build q knowing d (see
Annex B.3 for the definition of the polar form). Let us denote $G = (g_{ij})_{i,j}$ with
$1 \le i, j \le n$ the matrix of the polar form of q, computed from the distances $d(i, j)$
by

$$g_{ij} = -\frac{1}{2}\left(d_{ij}^2 - \Delta_i - \Delta_j + \Delta\right), \tag{9.5.2}$$

with

$$\Delta_i = \frac{1}{n}\sum_k d_{ik}^2, \qquad \Delta_j = \frac{1}{n}\sum_k d_{kj}^2, \qquad \Delta = \frac{1}{n^2}\sum_{k,\ell} d_{k\ell}^2.$$

G is a symmetric bilinear form. Let us assume for the sake of simplicity that it is
definite (the extension to the case where it is non-definite, i.e., is not full rank, is
straightforward). The quadratic form is given by

$$q(x) = G(x, x), \qquad \forall\, x \in E,$$

or, if $x = \sum_i x_i\, a_i$,

$$q(x) = \sum_{i,j=1}^{n} g_{ij}\, x_i x_j.$$

Now that we have the quadratic form, we need to construct the embedding φ.

Building the Quadratic Embedding   Let us now denote by (ω_k, z_k) the eigenpairs of G, with $\omega_k \in \mathbb{R}$ and $z_k \in E$, i.e., pairs (ω_k, z_k) with

$$Gz_k = \omega_k\, z_k, \tag{9.5.3}$$

or

$$G Z = Z\,\Omega.$$

(Z is the $n \times n$ matrix with eigenvectors as columns, and Ω is the $n \times n$ diagonal matrix with ω_k on the diagonal.) We have

$$G = Z\,\Omega\, Z^{\mathrm{T}} \tag{9.5.4}$$

(indeed, if $G' = Z\Omega Z^{\mathrm{T}}$, we have $G'Z = Z\Omega = GZ$ as $Z^{\mathrm{T}}Z = \mathbb{I}_n$, and as Z is invertible, $G' = G$). Let us separate the positive from the negative eigenvalues of G, i.e., denote

$$\omega_1 \geq \ldots \geq \omega_p > 0 > \omega'_1 \geq \ldots \geq \omega'_m, \tag{9.5.5}$$

(we assume for the sake of simplicity that 0 is not an eigenvalue of G) or, if (p, m) is the signature of q (see Appendix B.3)

$$\begin{cases} \omega_k > 0 \text{ if } k \leq p \\ \omega_k < 0 \text{ if } k > p \end{cases} \qquad \text{with} \qquad \omega'_j = \omega_{p+j}.$$

Let us denote

$$\omega_k = \sigma_k^2, \qquad \omega'_j = -\theta_j^2, \tag{9.5.6}$$

and

$$\begin{cases} G\,u_k = \sigma_k^2\, u_k \\ G\,v_j = -\theta_j^2\, v_j. \end{cases}$$

Let us use blockwise notations

$$Z = [U, V], \qquad \Omega = \begin{pmatrix} \Sigma^2 & 0 \\ 0 & -\Theta^2 \end{pmatrix}. \tag{9.5.7}$$

Then

$$\begin{cases} G\,U & = U\,\Sigma^2 \\ G\,V & = -V\,\Theta^2. \end{cases}$$

Rewriting Eq. (9.5.4) $G = Z\Omega Z^{\mathrm{T}}$ with this blockwise decomposition leads to

$$G = U\,\Sigma^2\,U^{\mathrm{T}} - V\,\Theta^2\,V^{\mathrm{T}}, \tag{9.5.8}$$

illustrated by

$$G = \begin{bmatrix} U & V \end{bmatrix} \begin{bmatrix} \Sigma^2 & 0 \\ 0 & -\Theta^2 \end{bmatrix} \begin{bmatrix} U^{\mathrm{T}} \\ V^{\mathrm{T}} \end{bmatrix} = \begin{bmatrix} U\Sigma^2 & -V\Theta^2 \end{bmatrix} \begin{bmatrix} U^{\mathrm{T}} \\ V^{\mathrm{T}} \end{bmatrix} = U\,\Sigma^2\,U^{\mathrm{T}} - V\,\Theta^2\,V^{\mathrm{T}}$$

Let us denote

$$\begin{cases} X & = U\,\Sigma \\ Y & = V\,\Theta. \end{cases} \tag{9.5.9}$$

Then

$$G = XX^{\mathrm{T}} - YY^{\mathrm{T}}. \tag{9.5.10}$$

One recovers in X the components obtained by clipping to 0 the eigenvalues ω'_j and complements this classical result with a second point cloud associated with the negative eigenvalues of the Gram matrix.

Algorithm This leads to the following algorithm (for clarity, here we denote by λ the positive eigenvalues of G and by ψ the negative ones, which were denoted by ω and ω' in the text).

Algorithm 13 Quadratic embedding of a discrete metric space

1: **input** $D \in \mathbb{R}^{n \times n}$: $D[i, j] = d_{ij}$
2: **compute** $\Delta_i = \frac{1}{n} \sum_k d_{ik}^2, \quad \Delta_j = \frac{1}{n} \sum_k d_{kj}^2, \quad \Delta = \frac{1}{n^2} \sum_{i,j} d_{ij}^2$
3: **compute** $G \in \mathbb{R}^{n \times n}$: $G[i, j] = -\frac{1}{2}\left(d_{ij}^2 - \Delta_i - \Delta_j + \Delta\right)$
4: **compute** (ω_k, z_k) : $G z_k = \omega_k z_k$
5: **denote** Λ, Ψ, diagonal matrices of eigenvalues $\lambda > 0$ and $\psi < 0$
6: **compute** $\Sigma = \Lambda^{1/2}, \Theta = (-\Psi)^{1/2}$
7: **denote** U, V, eigenvectors of G with $\lambda > 0$, and $\psi < 0$
8: **compute** $X = U\Sigma, Y = V\Theta$
9: **return** X, Y, Σ, Θ

Summary and Quality of the Low Rank Approximation   We have a metric space (M, d) with $|M| = n$. There exists a quadratic space (E, q) with $\dim E = n$ and a map

$$M \xrightarrow{\varphi} E$$

with $a_i = \varphi(i)$ such that

$$d^2(i, j) = q(a_i - a_j).$$

Let G be the polar form of q, i.e.,

$$g_{ij} = G(a_i, a_j) = \frac{1}{2}(q(a_i + a_j) - q(a_i) - q(a_j)).$$

It is called the kernel matrix of q and can be computed knowing the distances by

$$g_{ij} = -\frac{1}{2}\left(d_{ij}^2 - \Delta_i - \Delta_j + \Delta\right),$$

with

$$\Delta_i = \frac{1}{n} \sum_j d_{ij}^2, \qquad \Delta_j = \frac{1}{n} \sum_i d_{ij}^2, \qquad \Delta = \frac{1}{n^2} \sum_{i,j} d_{ij}^2.$$

The quadratic form q is defined by its polar form: $q(x) = G(x, x)$. The signature of q is (p, m) with p being the number of positive eigenvalues of G and m the number of negative eigenvalues. We denote by E_+ (resp. E_-) the subspace of E spanned by the eigenvectors of G associated with a positive (resp. negative) eigenvalue of G. Then, one can write

$$E = E_+ \oplus E_-$$

and, for any point $a_i \in E$,

$$a_i = x_i \oplus y_i, \qquad \text{with} \quad \begin{cases} x_i & \in E_+ \\ y_i & \in E_-. \end{cases}$$

Two point clouds X and Y are built with Algorithm 13, with X being made of n points in $\mathbb{R}^p$ and Y of n points in $\mathbb{R}^m$ such that

$$G = XX^\mathrm{T} - YY^\mathrm{T}.$$

As G is symmetric, its eigenvectors are orthogonal. The columns of X (respectively, Y) are the eigenvectors associated with positive (respectively, negative) eigenvalues of G. Then $X^\mathrm{T}Y = \mathbf{0}$. Hence

$$\begin{aligned} \|G\|^2 &= \|XX^\mathrm{T} - YY^\mathrm{T}\|^2 \\ &= \|XX^\mathrm{T}\|^2 + \|YY^\mathrm{T}\|^2 - 2\langle XX^\mathrm{T}, YY^\mathrm{T} \rangle \\ &= \|XX^\mathrm{T}\|^2 + \|YY^\mathrm{T}\|^2, \end{aligned} \tag{9.5.11}$$

because

$$\begin{aligned} \langle XX^\mathrm{T}, YY^\mathrm{T} \rangle &= \mathrm{Tr}\,\{XX^\mathrm{T}(YY^\mathrm{T})^\mathrm{T}\} \\ &= \mathrm{Tr}\,\{XX^\mathrm{T}\,YY^\mathrm{T}\} \\ &= \mathrm{Tr}\,\{X\,(X^\mathrm{T}Y)\,Y^\mathrm{T}\} \\ &= 0. \end{aligned} \tag{9.5.12}$$

If negative eigenvalues are clipped to 0, one has $G \approx XX^\mathrm{T}$, and the quality of representation of G by XX^T is $\|XX^t\|/\|G\|$. Similarly, knowing that

$$g_{ij} = \langle x_i, x_j \rangle - \langle y_i, y_j \rangle,$$

(this is another formulation of $G = XX^\mathrm{T} - YY^\mathrm{T}$), one has

$$\begin{aligned} d^2(i, j) &= q(a_i - a_j) \\ &= G(a_i - a_j, a_i - a_j) \\ &= G(a_i, a_i) + G(a_j, a_j) - 2G(a_i, a_j) \\ &= g_{ii} + g_{jj} - 2\,g_{ij} \\ &= \|x_i\|^2 - \|y_i\|^2 + \|x_j\|^2 - \|y_j\|^2 - 2\langle x_i, x_j \rangle + 2\langle y_i, y_j \rangle \\ &= \|x_i - x_j\|^2 - \|y_i - y_j\|^2. \end{aligned} \tag{9.5.13}$$

So, $\|y_i - y_j\|^2$ quantifies the discrepancy between $d^2(i, j)$ and $\|x_i - x_j\|^2$ in classical MDS (where negative eigenvalues and eigenvectors are clipped to 0) with full rank. This shows also that distances in the point cloud built by classical

MDS with clipping negative eigenvalues to 0 are overestimated as $\|x_i - x_j\|^2 = d_{ij}^2 + \|y_i - y_j\|^2 \geq d_{ij}^2$.

Notes and References See Appendix B.3 for a short presentation of quadratic forms and spaces. The classification of quadratic forms when $\mathbb{K} = \mathbb{R}$ or $\mathbb{C}$ depends on the matrix of the polar form and is well understood and given, for example, in [dSP10]. The classification on an arbitrary field is said to be immensely difficult [Ber87]. Embedding from a metric space into a quadratic space has been studied, e.g., in [Gol84, Gow85], and is presented in [PD05, section 3.5] where it is called *pseudo-Euclidean embedding*.

Appendix A
References in Textbooks

Canonical Correlation Analysis	[And58]	chapter 12
	[Rao73]	section 8f1
	[MKB79]	chapter 10
	[Jol02]	section 9.3
	[Ize08]	section 7.3
	[HTF09]	section 14.5.1
	[Mur12]	section 12.5.3
Correspondence Analysis	[Gre84]	the book
	[Ize08]	chapter 17
Factor Analysis	[And58]	section 14.7
	[Rao73]	section 8f4
	[MKB79]	chapter 9
	[Ize08]	section 15.4
	[HTF09]	section 14.7.1
	[Mur12]	section 12.1
Independent Component Analysis	[Ize08]	section 15.3
	[HTF09]	section 14.7.2
Karhunen Loeve transform	[Mur12]	section 12.2, p. 387
Latent variables	[Bis06]	chapter 12
	[Ize08]	chapter 15
	[HTF09]	section 14.7.1
	[Mur12]	chapter 12
Multiple Correspondence Analysis	[Ize08]	section 17.4

A. Franc, *Linear Dimensionality Reduction*, Lecture Notes in Statistics 228,
https://doi.org/10.1007/978-3-031-95785-7

Principal Component Analysis	[And58]	chapter 11
	[MKB79]	chapter 8
	[Jol02]	the book
	[Bis06]	section 12.1
	[Ize08]	section 7.2
	[Mur12]	section 12.2
Probabilistic PCA	[Bis06]	
	[Mur12]	section 12.2.4
Sensible PCA	[Mur12]	section 12.2, p. 387
Sparse PCA	[HTF09]	section 14.5.5
Supervised PCA	[Mur12]	section 12.5.1
Classical Scaling	[CC01]	section 2.2
	[Ize08]	section 13.6
Multidimensional Scaling	[CC01]	the book
	[MKB79]	chapter 14
	[Ize08]	chapter 13
	[HTF09]	section 14.8
Non metric scaling	[CC01]	chapter 3
	[Ize08]	section 13.9

Appendix B

In this appendix we gather some elementary facts and notions that will be useful for deriving or understanding the methods presented in this book. The material that appears here should be considered as a reminder or as a memory aid. It is organized as follows.

Some basic notions of linear algebra are recalled in Appendix B.1, with a focus on eigenvalues and their perturbations.

PCA with metrics deriving from a Euclidean structure is a common framework for presenting Correspondence Analysis, Canonical Analysis, and Multiple Canonical Analysis. The notion of inner product, associated norm and distance (useful for approximation of matrices), and projection in a space embedded with a Euclidean structure are presented in Appendix B.2.

Elementary notions defining a quadratic space and a pseudo-Euclidean structure are recalled in Appendix B.3. This will be used for MDS.

Several calculations in this book are easier to follow using tensor-based notations, which are however not compulsory. A primer in multilinear algebra is presented in Appendix B.4, which should be considered as a crash course or a reference card for tensor-based notations, especially for the SVD, in linear algebra.

Random projection for deriving random SVD can be considered as the guiding principle of this book, or at least its "raison d'être," as it makes it possible to scale the methods presented just by scaling SVD, which is a key step common to all methods. These notions are presented in Appendix B.5, used especially in Chap. 2 (Randomized SVD), and are behind the scenes of all methods that use SVD. This is the only appendix which is technical and not a summary of elementary facts.

A. Franc, *Linear Dimensionality Reduction*, Lecture Notes in Statistics 228,
https://doi.org/10.1007/978-3-031-95785-7

B.1 Preliminaries in Linear Algebra

Let us recall here some basic facts in linear algebra. Linear algebra can have an abstract setting, with the notions of vector spaces and linear maps, or numerical setting, working with arrays (vectors and matrices). If E, F are two finite dimensional vector spaces, a matrix A is the expression of a linear map $E \longrightarrow F$ by an array once bases have been chosen in E and F.

Vector Space and Linear Map

$\diamond$ Let $\mathbb{K}$ be a field, usually $\mathbb{R}$ or $\mathbb{C}$, and here $\mathbb{R}$ unless otherwise stated. A vector space E over $\mathbb{K}$ is a set endowed with two operations:

$\rightarrow$ An addition, $+$, such that $(E, +)$ is an Abelian group
$\rightarrow$ A multiplication by a scalar

$$\mathbb{K} \times E \longrightarrow E$$

$$(\alpha, u) \longrightarrow \alpha u.$$

Multiplication by a scalar verifies the following properties:

$$\begin{vmatrix} \alpha(u + v) = \alpha u + \alpha v \\ (\alpha + \beta)u = \alpha u + \beta u \\ \alpha(\beta u) \quad = \quad (\alpha\beta)u \\ 1u \quad = \quad u. \end{vmatrix}$$

$\diamond$ A vector subspace of E is a subset of E which is closed under the addition and the multiplication by a scalar. It is itself a vector space on $\mathbb{K}$.
$\diamond$ A basis of a vector space E is a collection $(u_k)_k$ of vectors $u_k \in E$ such that any vector $u \in E$ can be written as

$$u = \sum_k \alpha_k u_k,$$

in a unique manner where $\alpha_k \in \mathbb{K}$. Every vector space has a basis. The proof of this theorem uses Zorn lemma, which is equivalent to the axiom of choice. Conversely, it is possible to show that if every vector space has a basis, then Zorn lemma is verified. All bases have the same cardinality. This cardinality is the dimension of the vector space. Usually, the dimension is either an integer $n \in \mathbb{N}$ or $\aleph_0$, the cardinality of $\mathbb{N}$ as in many vector spaces in functional analysis. However, more complex situations can be met, like $\mathbb{R}$ as a vector space on $\mathbb{Q}$. The vector spaces we will work with in these notes are always finite dimensional.

⋄ Let E, F be two vector spaces on the same field $\mathbb{K}$. A linear map

$$E \xrightarrow{\ L\ } F$$

from E on F is a map which preserves the linear structure, i.e.,

$$
\begin{aligned}
L(u + v) &= Lu + Lv \\
L(\alpha u) &= \alpha\, Lu.
\end{aligned}
$$

(It is customary for linear maps to write Lu instead of $L(u)$ when there is no ambiguity.) The image of a linear subspace $E' \subset E$ is a vector subspace $F' \subset F$. The preimage of a vector subspace $F' \subset F$ is a vector subspace $E' \subset E$.

⋄ Let $n = \dim E \in \mathbb{N}$ and $m = \dim F \in \mathbb{N}$. Let $(u_j)_j$ with $1 \le j \le n$ be a basis of E and $(v_i)_i$ with $1 \le i \le m$ a basis in F. Let

$$Lu_j = \sum_i \alpha_{ij} v_i.$$

Then, the matrix $A = (\alpha_{ij})_{i,j} \in \mathbb{R}^{m \times n}$ is the matrix of the linear map L in basis $(u_j)_j$ for E and $(v_i)_i$ for F. We have, for any vector $u = \sum_j \beta_j u_j \in E$,

$$
\begin{aligned}
Lu = L\left(\sum_j \beta_j u_j \right) \\[2mm]
= \sum_j \beta_j Lu_j \\[2mm]
= \sum_j \beta_j \left(\sum_i \alpha_{ij} v_i \right) \\[2mm]
= \sum_i \left(\sum_j \alpha_{ij} \beta_j \right) v_i.
\end{aligned}
\qquad\qquad \text{(B.1.1)}
$$

Hence, the j-th column of A is the vector $Lu_j \in \mathbb{R}^m$ with coordinates $(\sum_{\ell=1}^n \alpha_{i\ell} \beta_\ell)_{1 \le i \le m}$.

⋄ The kernel of L is the set of vectors $u \in E$ the image of which is $0 \in F$

$$\ker L = \{ u \in E \mid Lu = 0 \}. \qquad\qquad \text{(B.1.2)}$$

It is a vector subspace of E.

⋄ The image of L is the set of vectors $v \in F$ which have a preimage in E

$$\operatorname{im} L = \{ v \in F \mid \exists\, u \in E \quad \text{s.t.} \quad v = Lu \}. \qquad\qquad \text{(B.1.3)}$$

It is a vector subspace of F.

$\diamond$ The rank nullity theorem states that

$$\dim \ker L + \dim \operatorname{im} L = \dim E. \tag{B.1.4}$$

$\diamond$ A linear map from E to E is called an endomorphism. If it is an isomorphism, it is called an automorphism. The set of all endomorphisms of E, denoted end E, is an associative algebra for the operation $+$ and multiplication by a scalar for the vector space structure and the composition $\circ$ for it to be an algebra.

Eigenspace, Eigenvector, and Eigenvalue

$\diamond$ Let E be a finite dimensional vector space on a field $\mathbb{K}$ and L an endomorphism in E

$$E \xrightarrow{\ \ L\ \ } E.$$

An eigenvector of L is a vector $u \neq 0 \in E$ such that

$$Lu = \lambda u, \tag{B.1.5}$$

for some $\lambda \in \mathbb{K}$. λ is called an eigenvalue of L. If A is the matrix of L in a given basis, one writes also $Ax = \lambda x$ if x is the expression of u in the same basis, and (x, λ) is an eigenpair of A. The set of eigenvalues of A is called the spectrum of A:

$$\operatorname{Sp} A = \{\lambda \in \mathbb{K} \mid \exists\, x \neq 0 \quad \text{s.t.} \quad Ax = \lambda x\}.$$

The eigenspace of A associated with eigenvalue λ is the set of all eigenvectors associated with λ. This space completed with $\{0\}$ is the kernel of $A - \lambda \mathbb{I}$. In order to avoid technicalities, one includes $\{0\}$ in the eigenspace associated with an existing eigenvalue. The eigenspace is then $\ker (A - \lambda \mathbb{I})$ and is a linear subspace of E.

$\diamond$ Even if matrix A is real, its eigenvalues can be complex. For example, if

$$A = \begin{pmatrix} 0 & 1 \\ -1 & 0 \end{pmatrix}, \qquad x = \begin{pmatrix} y \\ z \end{pmatrix},$$

we have

$$Ax = \begin{pmatrix} z \\ -y \end{pmatrix}.$$

If $Ax = \lambda x$,

$$\begin{cases} z & = \lambda y \\ -y & = \lambda z, \end{cases}$$

hence $z = -\lambda^2 z$, so $\lambda^2 = -1$ (we cannot have $z = 0$ because if so, $y = z = 0$ and $x = 0$) and $\lambda = \pm i \in \mathbb{C}$. The eigenvalues of A, if they exist, are the root of the characteristic polynomial of A

$$\lambda \in \mathrm{Sp}\, A \quad \Longrightarrow \quad \det(A - \lambda \mathbb{I}) = 0, \tag{B.1.6}$$

where $\mathbb{I}$ is the identity matrix. The roots of a real polynomial can be complex.

$\diamond$ A square matrix A acting on E is diagonalizable if E has a basis of eigenvectors of A. If $A \in \mathbb{K}^{n \times n}$, the sum of the dimensions of its eigenspaces is n. Not all matrices are diagonalizable. For example, if

$$A = \begin{pmatrix} 0 & 1 \\ 0 & 0 \end{pmatrix}, \qquad x = \begin{pmatrix} y \\ z \end{pmatrix},$$

and $Ax = \lambda x$ leads to

$$\begin{cases} 0.y + z & = \lambda y \\ 0.y + 0.z & = \lambda z, \end{cases}$$

then $\lambda = 0$, $z = 0$, and $y \in \mathbb{K}$. The vector space spanned by the eigenvectors of A has dimension 1, and A is not diagonalizable. A matrix that is not diagonalizable is called defective. A nilpotent matrix is a matrix A such that there exists an integer $m > 0$ with $A^m = 0$. A nilpotent matrix is defective (except the zero matrix $\mathbf{0}$, because any vector of E is an eigenvector of $\mathbf{0}$ associated with eigenvalue $\lambda = 0$).

$\diamond$ A matrix A is diagonalizable, or non-defective, if there exists an invertible matrix P such that $A = P \Lambda P^{-1}$, where Λ is a diagonal matrix (all matrices in this formula are in $\mathbb{K}^{n \times n}$). $A = P \Lambda P^{-1}$ is called the eigen-decomposition of A. The columns of P are eigenvectors of A. It can be seen through $AP = P\Lambda$. If A is real symmetric, its eigen-decomposition exists, with P orthogonal ($P^{-1} = P^{\mathrm{T}}$), and its spectrum is real.

Notes and References There exist many excellent books as overview or in-depth studies of Linear Algebra and Matrix Calculus, some dedicated to applications in statistics, among which are [Sch97, HJ12, Str19, GM19, GM20].

Perturbation of Eigenvalues

A matrix A being given, there are two sources of errors when computing its eigenvalues:

→ A numerical error while using rounding during the calculation: This is addressed by numerical analysis of eigen-decomposition.

→ When the matrix is the outcome of an experiment, i.e., a dataset, the data can be corrupted, which leads to a corruption of the eigenvalues as well.

In this section, we address the second source of errors: possible corruption of a dataset (which is dealing with uncertainty rather than error). It will be referred to as a perturbation. The theory of perturbations of the eigenvalues of a given matrix is the study of the variations of the eigenvalues of a given matrix under a perturbation of its coefficients.

We know by the Rouché theorem that the eigenvalues of a matrix A are continuous functions of the coefficients of A. Pointedly speaking, let $A, H \in \mathbb{K}^{n \times n}$, where $\mathbb{K}$ is $\mathbb{R}$ or $\mathbb{C}$, and $\epsilon \in \mathbb{R}$. What can be said on the localization of eigenvalues of $A + \epsilon H$ knowing the spectrum of A? A first observation is that the eigenvalues of A are the roots of a polynomial (the characteristic polynomial, of degree n if $\dim A = n$, $\det(A - \lambda \mathbb{I}) = 0$). As, in general, the roots of a polynomial as functions of its coefficients are unstable, it is likely that the eigenvalues of a matrix are unstable. This will be shown, but a good news is that the eigenvalues of a symmetric matrix are stable under a perturbation by a symmetric matrix, i.e., vary with the same order of magnitude as the coefficients of the matrix. This is Weyl's theorem which dates back to 1912.

Let us have (this example is borrowed from [SS90, chapter IV])

$$A = \begin{pmatrix} 0 & 1 & 0 & 0 \\ 0 & 0 & 1 & 0 \\ 0 & 0 & 0 & 1 \\ 0 & 0 & 0 & 0 \end{pmatrix}.$$

It is easy to show that $\operatorname{Sp} A = \{0\}$. Let us now have

$$A + \epsilon H = \begin{pmatrix} 0 & 1 & 0 & 0 \\ 0 & 0 & 1 & 0 \\ 0 & 0 & 0 & 1 \\ \epsilon & 0 & 0 & 0 \end{pmatrix}.$$

Then,

$$\operatorname{Sp}(A + \epsilon H) = \{\pm \epsilon^{1/4}, \pm i \epsilon^{1/4}\}. \tag{B.1.7}$$

This can be generalized to any dimension n, which shows that the perturbation of eigenvalues can be in $\epsilon^{1/n}$ if $A \in \mathbb{R}^{n \times n}$. If $n = 100$ and $\epsilon = 10^{-2}$, then $\epsilon^{1/n} \approx 0.95$. If ϵ' is the magnitude of the perturbation of the eigenvalues, we have $\epsilon'/\epsilon = 0.95/10^{-2} = 95$. The perturbation is amplified about 100 times.

The first tool needed is a way to compare the spectrum of the initial and the perturbed matrix. The tools therefore are the spectral variation and distances, like Hausdorff distance and matching distance between spectra. They are presented hereafter. Then, some bounds are given on distances between spectra of A and B, and used with $B = A + \epsilon H$ being a perturbatuion of A. This is a key question in numerical analysis which has been thoroughly studied (see the Notes and References section).

Spectral Variation Let $A, B \in \mathbb{R}^{n \times n}$ or $\mathbb{C}^{n \times n}$ with $\mathrm{Sp}A = \{\lambda_k\}_k$ and $\mathrm{Sp}B = \{\mu_k\}_k$ with $1 \leq k \leq n$. Then, the spectral variation of B relatively to A is the number

$$\mathrm{sv}_A(B) = \max_i \left(\min_j |\lambda_i - \mu_j| \right).$$

It measures the maximum possible gap between an eigenvalue in A and the closest eigenvalue in B.

Matching Distance The spectral variation is not a distance. However, a distance between spectra can be built as

$$d_H(\mathrm{Sp}\, A, \mathrm{Sp}\, B) = \max\{\mathrm{sv}_A(B),\, \mathrm{sv}_B(A)\}.$$

It is the Hausdorff distance between the spectra of A and B. However, this distance is not term-by-term comparison of eigenvalues. Hence the matching distance is defined as

$$\mathrm{md}\,(\mathrm{Sp}\, A, \mathrm{Sp}\, B) = \min_\pi \left(\max_i |\lambda_i - \mu_{\pi(i)}| \right), \tag{B.1.8}$$

where π runs over all permutations of $\mathrm{Sp}\, B$.

Ostrowski and Elsener Theorem It can be shown that

$$\mathrm{md}\,(\mathrm{Sp}\, A, \mathrm{Sp}\, B) \leq 4 \times 2^{-1/n} \left(\|A\| + \|B\| \right)^{1-1/n} \|A - B\|^{1/n}, \tag{B.1.9}$$

or

$$\mathrm{md}\,(\mathrm{Sp}\, A, \mathrm{Sp}\, B) \leq 4 \times 2^{-1/n} \left(\|A\| + \|B\| \right) \left(\frac{\|A - B\|}{\|A\| + \|B\|} \right)^{1/n}, \tag{B.1.10}$$

where

$$\|X\| = \sqrt{\lambda_{\max}(X^*X)} = \max_{\|u\|=1} Xu. \tag{B.1.11}$$

Hence, if $B = A + \epsilon H$, this yields

$$\mathrm{md}\,(\mathrm{Sp}\,A, \mathrm{Sp}\,A + \epsilon H) \le 4 \times 2^{-1/n}\,(\|A\| + \|A + \epsilon H\|)\left(\frac{\|\epsilon H\|}{\|A\| + \|A + \epsilon H\|}\right)^{1/n}$$

$$\le C\epsilon^{1/n}. \tag{B.1.12}$$

The bound $h = 1/n$ in ϵ^h is reached in the example.

Henrici Theorem There exists a general and powerful result for the localization of eigenvalues of a matrix which is the sum of two symmetric matrices. Let A, B, $C \in \mathbb{R}^{n \times n}$, symmetric, with

$$B = A + C. \tag{B.1.13}$$

Let us denote

$$\begin{cases} \mathrm{Sp}\,A = \{\alpha_i\} \\ \mathrm{Sp}\,B = \{\beta_j\} \\ \mathrm{Sp}\,C = \{\gamma_k\}. \end{cases} \tag{B.1.14}$$

Let us assume that

$$\begin{cases} \alpha_1 \ge \ldots \ge \alpha_n \\ \beta_1 \ge \ldots \ge \beta_n \\ \gamma_1 \ge \ldots \ge \gamma_n. \end{cases} \tag{B.1.15}$$

Then (recall that $C = B - A$)

$$\forall\, i, \quad \gamma_n \le \beta_i - \alpha_i \le \gamma_1. \tag{B.1.16}$$

This theorem has a nice consequence for symmetric perturbation of eigenvalues of a symmetric matrix. Let us assume that $C = \epsilon H$ or $B = A + \epsilon H$. Let us denote

$$\overline{h} = \sup_{i,j} |h_{ij}|. \tag{B.1.17}$$

Then

$$\forall\, i, \quad |\gamma_i| \le \epsilon \overline{h}, \tag{B.1.18}$$

and

$$\forall\, i, \quad |\beta_i - \alpha_i| \leq \epsilon \overline{h}. \tag{B.1.19}$$

Hence the response of any eigenvalue of a symmetric matrix to a symmetric perturbation is bounded by a term linear with the perturbation, in $O(\epsilon)$, and not in $O(\epsilon^{1/n})$ as for any matrix. Eigenvalues of symmetric matrices are much more stable under symmetric perturbations.

Application to PCA Let us now have a matrix $A \in \mathbb{R}^{n \times p}$. PCA involves the computations of eigenvalues and eigenvectors of $A' = A^T A$. Let us have a small perturbation of A: $A \longrightarrow A + \epsilon H$. As a consequence, the perturbation C' of A' is given by

$$\begin{aligned} A' + C' &= (A^T + \epsilon H^T)(A + \epsilon H) \\ &= A' + \epsilon(H^T A + A^T H) + \epsilon^2\, H^T H \\ &= A' + \epsilon(H^T A + A^T H + \epsilon\, H^T H) \end{aligned} \tag{B.1.20}$$

with $H^T A + A^T H + \epsilon\, H^T H$ being symmetric. Then, C' is a symmetric perturbation, and the Henrici theorem applies. The variation of the eigenvalues is bounded by a term linear with ϵ.

This can be derived directly from Weyl's 1912 theorem (see the Notes and References section), knowing that $\forall\, i, \quad |\lambda_i| \leq \sup_{i,j} |\alpha_{ij}|.$

Application to SVD PCA relies also on the SVD of A, written as

$$A = U \Sigma V^T.$$

We have

$$A^T A = V \Sigma^2 V^T,$$

and Σ^2 is the diagonal matrix of eigenvalues of $A^T A$. Then, the above observation on the response of the eigenvalues of $A^T A$ to a perturbation of A applies. The error on σ^2 is in $O(\epsilon)$, so on σ in $O(\epsilon)$ as well. There is however a difficulty for the stability of small singular values. Indeed, for a singular value σ, the perturbation is

$$\sigma^2 \longrightarrow \sigma'^2 = \sigma^2 + h\epsilon$$

if the perturbation of A is ϵH, $h \leq \max_{i,j} |h_{ij}|$ and $\epsilon \ll h$. Then, as $\sqrt{a^2 + \epsilon} \simeq a\left(1 + \frac{\epsilon}{2a}\right)$, we have

$$\sigma' \simeq \sigma\left(1 + \frac{\epsilon h}{2\sigma}\right), \tag{B.1.21}$$

and, if $\sigma \ll 1$, this is unstable.

Notes and References Here, we have followed [SS90, chapter IV], with historical notes p. 176 & sq. Perturbation theory of eigenvalues of matrices or linear operators has a long history, from a result by H. Weyl in 1912. It states that if A, B are self-adjoint matrices (a self-adjoint matrix is a matrix $A \in \mathbb{C}^{n \times n}$ with $A = A^*$, where $A^* = \overline{A^{\mathsf{T}}}$) with spectra Sp $A = \{\alpha_i\}$ and Sp $B = \{\beta_j\}$, with $\alpha_1 \geq \ldots \geq \alpha_n$ and $\beta_1 \geq \ldots \geq \beta_n$, then $\max_k |\alpha_k - \beta_k| \leq \|A - B\|_{\mathrm{sp}}$, where the norm is the spectral norm: $\|A\|_{\mathrm{sp}} = \max_{\|x\|=1} \|Ax\|$. Several decades or efforts have aimed at finding similar bounds in more general situations, i.e., non-self-adjoint matrices or other norms. Much work has concerned so-called normal matrices (a normal matrix is a matrix A such that $AA^* = A^*A$). A matrix is normal if, and only if, it is diagonal in some orthogonal basis. As eigenvalues of normal matrices can be complex, they cannot be ordered as in $\mathbb{R}$. The notion of matching distance permits to compare spectra with complex values. The question to know whether md (Sp A, Sp B) $\leq \|A - B\|$ for normal matrices have been object of intensive researches for decades. Hoffman and Wielandt have proved in 1953 (40 years after Weyl's result) a similar result for normal matrices, but with Frobenius norm $\|A\|_F = \sqrt{\mathrm{Tr}\, A^*A} = \left(\sum_{i,j} |\alpha_{ij}|^2 \right)^{1/2}$: md_F (Sp A, Sp B) $\leq \|A - B\|_F$, where md_F is a matching distance adapted to Frobenius norm (ℓ^2 norm). See Bathia [Bha07] and [SS90] for historical notes, from which those few milestones have been borrowed.

Due to its role in numerical analysis, spectral variation has been thoroughly studied over several decades (see, e.g., Henrici [Hen62], Bhatia [Bha82], and Elsner [Els82]). Classical books on bounds for eigenvalue perturbation theory are [Bha87, SS90]. The inequality (B.1.9) appears in Bhatia [BEK90]. Several recent results for upper bounds of matching distance between two spectra appear in Galantái [GH08]. Ostrowski theorem dates from 1940 and has been published in Ostrowski, A. (1940) Recherches sur la méthode de Gräffe et les zéros des polynômes et des series de Laurent. *Acta Math.*, **72**:99–257 (see [Hol92]).

B.2 Euclidean Vector Space

In this appendix, we recall the main elementary results about the Euclidean structure of a finite dimensional real vector space E. An inner product in E is defined as a symmetric, definite, positive bilinear form. It induces a norm (then, E is a normed space) which in turn induces a distance on E (hence E is a metric space). The next step is to define a topology from the distance, but this will not be necessary here. E being a metric space, the projection of a vector $x \in E$ on a subspace $F \subset E$ is the vector $y \in F$ such that $d(x, y)$ is minimal. The computation of y knowing x and an inner product in E is given.

Inner Product and Euclidean Structure

Let $E = \mathbb{R}^n$ be a finite dimensional real vector space. A bilinear form B on E is a map

$$E \times E \xrightarrow{\ \ B\ \ } \mathbb{R},$$

which is linear on each component, i.e., $B(x, .) : y \mapsto B(x, y)$ and $B(., y) :\, x \mapsto B(x, y)$ are linear. A bilinear form is:

◇ Symmetric if $B(y, x) = B(x, y)$
◇ Positive if for any x, $B(x, x) \geq 0$
◇ Definite if $B(x, x) = 0$ implies that $x = 0$

A symmetric definite positive bilinear form is called an inner product, and a vector space E equipped with such an inner product is called a Euclidean space. The inner product is classically denoted by

$$B(x, y) = \langle x, y \rangle.$$

The map

$$E \xrightarrow{\ \ \|\cdot\|\ \ } \mathbb{R}^+,$$

defined by

$$\|x\| = \sqrt{\langle x, x \rangle},$$

is a norm. It is called the ℓ^2 norm or, when E is a space of matrices, like $R^{n \times p}$, the Frobenius norm.

It induces on E a structure of metric space (E, d) with

$$d(x, y) = \|x - y\|.$$

It is easy to show that d is a distance.

Two vectors $x, y \in E$ are said orthogonal if $\langle x, y \rangle = 0$. In such a case, we have Pythagoras theorem:

$$\|x + y\|^2 = \|x\|^2 + \|y\|^2.$$

An orthonormal basis $(e_i)_i$ of E is a basis of E such that $\forall i, j, \langle e_i, e_j \rangle = \delta_i^j$ where δ_i^j is the Kronecker symbol.

Specification of a Euclidean Structure

Let $E = \mathbb{R}^p$, equipped with an inner product, called here the canonical inner product. It can be given by selecting a basis $\mathscr{E} = (e_1, \ldots, e_p)$ and setting it as an orthonormal basis. It is referred to as the canonical basis. A new inner product in $\mathbb{R}^p$ can be defined by an SDP matrix $P \in \mathbb{R}^{p \times p}$, and denoted $\langle .. \rangle_{\mathrm{P}}$, by

$$\langle x, y \rangle_{\mathrm{P}} = \langle x, Py \rangle. \tag{B.2.1}$$

As P is symmetric, we have $P = P^{\mathrm{T}}$, and $\langle x, y \rangle_{\mathrm{P}} = \langle Px, y \rangle$ as well. This means, component-wise in the canonical basis, that

$$\langle x, y \rangle_{\mathrm{P}} = \sum_{i,j=1}^{p} p_{ij} \, x_i \, y_j, \qquad p_{ji} = p_{ij}, \tag{B.2.2}$$

if $x = (x_i)_i$, $y = (y_j)_j$ and $P = (p_{ij})_{i,j}$. If $P = \mathbb{I}_p$, canonical inner product is recovered, as $p_{ij} = \delta_i^j$.

A case worth being studied in detail is when P is diagonal, i.e., $P = \operatorname{diag} w$ with $w = (w_1, \ldots, w_p)$, or

$$p_{ij} = \begin{cases} w_i & \text{if } i = j \\ 0 & \text{if } i \neq j. \end{cases}$$

In such a case

$$\langle x, y \rangle_w = \sum_{i=1}^{p} w_i \, x_i \, y_i, \tag{B.2.3}$$

and the norm $\|.\|_w$ is defined as

$$\|x\|_w^2 = \sum_{i} w_i x_i^2. \tag{B.2.4}$$

As P is SDP, there exists a unique SDP matrix Q such that $P = Q^2$. Then

$$\langle x, y \rangle_{\mathrm{P}} = \langle Qx, Qy \rangle, \tag{B.2.5}$$

and

$$\|x\|_{\mathrm{P}} = \|Qx\|. \tag{B.2.6}$$

Matrix Q can be computed as follows: If the SVD of P is $P = U\Sigma U^{\mathrm{T}}$, Q is given by $Q = U\Sigma^{1/2}U^{\mathrm{T}}$, as $Q^2 = U\Sigma^{1/2}U^{\mathrm{T}}U\Sigma^{1/2}U^{\mathrm{T}} = U\Sigma U^{\mathrm{T}} = P$, and Q is SDP. Unicity follows from unicity of SVD.

Remark One may wonder whether the metric should be given by P or Q. It is tempting to give it by Q, because Q establishes an isometry between $(\mathbb{R}^p, Q)$ and $(\mathbb{R}^p, \mathbb{I})$ by

$$(\mathbb{R}^p, Q) \longrightarrow (\mathbb{R}^p, \mathbb{I})$$
$$x \longrightarrow Qx,$$

as

$$\|x\|_Q = \|Qx\|.$$

However, in data analysis, it is classical to use weights, i.e., define metrics with diagonal matrices. Let

$$w = (w_1, \ldots, w_p) \in \mathbb{R}^{p+}.$$

Then, a distance between $x, x' \in \mathbb{R}^p$ with weights w is given by

$$d_w(x, x') = \|x - x'\|_w = \sqrt{\|x - x'\|_w^2} = \sqrt{\sum_i w_i (x_i - x'_i)^2}.$$

This is consistent with an isometry defined by

$$Q = \mathrm{diag}\left(\sqrt{w_1}, \ldots, \sqrt{w_p}\right),$$

as

$$d_w(x, x') = \sqrt{\sum_i \left(\sqrt{w_i} x_i - \sqrt{w_i} x'_i\right)^2}.$$

It is customary to define the weights by w, which are the diagonal elements of P, and not $\sqrt{w}$. Hence, it is consistent with this classical approach to define the metric by P, hence denote $\langle x, x' \rangle_P$. We will use P or Q indifferently, knowing that $P = Q^2$.

Projection Operator with Metrics

We define here the projection operator in $\mathbb{R}^p$ endowed with metrics defined by P. Let $v \in \mathbb{R}^p$ with $\|v\|_P = 1$. Let us denote by $\mathcal{P}_v$ the projection operator in $\mathbb{R}^p$ on $\mathbb{R}v$:

$$\mathbb{R}^p \xrightarrow{\mathcal{P}_v} \mathbb{R}v$$
$$x \longrightarrow \lambda v.$$

$\mathcal{P}_v x$ is defined as the vector λv such that $\|x - \lambda v\|_P$ is minimal. We have

$$\|x - \lambda v\|_P^2 = \|x\|_P^2 + \lambda^2 \|v\|_p^2 - 2\lambda \langle x, v \rangle_P$$
$$= \|x\|_P^2 + \lambda^2 - 2\lambda \langle x, v \rangle_P \qquad (\|v\|_P = 1).$$

As $\|x\|_P$ is fixed, this yields $\lambda^2 - 2\lambda \langle x, v \rangle_P$ is minimal, or, as the unknown is λ,

$$\lambda = \langle x, v \rangle_P = \langle x, Pv \rangle. \tag{B.2.7}$$

Then

$$\mathcal{P}_v x = \lambda v = \langle x, Pv \rangle \, v = (v \otimes Pv)x, \tag{B.2.8}$$

and

$$\mathcal{P}_v = v \otimes Pv. \tag{B.2.9}$$

Let us note that $\mathcal{P}_v \neq Qv \otimes Qv$ (indeed, $(Qv \otimes Qv)x = \langle Qv, x \rangle Qv \in \mathbb{R}Qv \notin \mathbb{R}v$). For a general subspace

$$F \subset \mathbb{R}^p = \operatorname{span}(v_1, \ldots, v_m)$$

with $(v_i)_i$ a basis of F orthonormal for P, the same calculation leads to

$$\mathcal{P}_F = \sum_{i=1}^{m} v_i \otimes Pv_i. \tag{B.2.10}$$

B.3 Quadratic Forms

E is a finite dimensional real vector space.

Quadratic and Polar Forms

Let B be a symmetric bilinear form on a finite dimensional real vector space E. Then, the map

$$E \xrightarrow{\ q\ } \mathbb{R},$$

defined by

$$q(x) = B(x, x),$$

is called a quadratic form. It is possible to compute B, knowing q, by

$$B(x, y) = \frac{1}{2}(q(x + y) - q(x) - q(y)) = \frac{1}{4}(q(x + y) - q(x - y)).$$

B is called the polar form of the quadratic form q.

Signature of a Quadratic Form

Let q be a quadratic form on E with $\dim E = n$. There exists a basis in E in which the matrix of the polar form B of q is

$$B = \begin{pmatrix} \mathbb{I}_p & 0 & 0 \\ 0 & -\mathbb{I}_m & 0 \\ 0 & 0 & 0 \end{pmatrix}. \tag{B.3.1}$$

The pair (p, m) is called the signature of the quadratic form and does not depend on the basis (Sylvester law of inertia). It is classical to denote $\mathbb{R}^n_{p,m}$ or $\mathbb{R}_{p,m}$ the space $\mathbb{R}^n$ equipped with the quadratic form

$$q(x) = \sum_{i=1}^{p} x_i^2 - \sum_{j=p+1}^{n'} x_j^2,$$

with $n' = p + m \leq n$.

Geometry in Quadratic Spaces

Euclidean spaces are those for which $p = n$ (and $m = 0$). Non-Euclidean quadratic spaces ($p < n$) are called pseudo-Euclidean spaces. In a Euclidean space, the map

$$(x, y) \longrightarrow d(x, y) = \sqrt{q(x - y)}$$

defines a distance. It is no longer the case in a pseudo-Euclidean space. For example, if $n = 2, p = m = 1$, the pseudo-sphere of radius 1 is defined by $x^2 - y^2 = 1$ and is a hyperbola (hence the name hyperbolic geometry for geometry in pseudo-Euclidean spaces).

See [Gar11] for further details on quadratic forms. Hyperbolic geometry is developed, for example, in [Ive92].

B.4   Primer in Multilinear Algebra

In this section, we denote

$$E = \mathbb{R}^n$$
$$F = \mathbb{R}^p$$
$$a, x \in E$$
$$b, y \in F$$
$$A \in \mathcal{L}(F, E)$$

E and F are each endowed with a Euclidean structure, denoted $\langle ., . \rangle$ for both.

If E is a finite dimensional real vector space, one can define E^*, the dual space of E, which is the vector space of linear forms on E. E and E^* are isomorphic, but there is no canonical isomorphism between both. However, it is equivalent to choose an isomorphism between E and E^* and to choose a Euclidean structure on E. Choosing a Euclidean structure is choosing a basis and setting that it is orthonormal. Then, if $a \in E$, the map denoted a^*

$$E \xrightarrow{\;\;a^*\;\;} \mathbb{R},$$

defined by

$$a^*(x) = \langle a, x \rangle,$$

is linear. The map

$$a \longrightarrow a^*$$

is an isomorphism between E and E^*. The same relations show that if an isomorphism $a \mapsto a^*$ is given between E and E^*, then the bilinear form B on $E \times E$ defined by

$$B(a, x) = a^*(x)$$

is an inner product.

Multilinear Algebra   It is convenient to make a small abstract detour through multilinear algebra to develop some calculations in linear algebra. The basic object of interest in multilinear algebra is a tensor, denoted T, which is formally a multilinear form on a Cartesian product of d vector spaces, $E_1 \times \ldots \times E_d$, considered here as finite dimensional. The integer d is called the order of the tensor, and the tuple $(n_1, \ldots, n_d)$, where $n_i = \dim E_i$ is its dimension. For example, for $d = 2$, we have a bilinear form

$$T \, : \, E_1 \times E_2 \longrightarrow \quad \mathbb{R}$$

$$(x_1, x_2) \quad \longrightarrow \quad T(x_1, x_2).$$

A tensor is also a multidimensional array (a d-dimensional one) when expressed in a given basis for each vector space. From now on in this section, we will restrict ourselves to bilinear forms, which can be expressed by matrices and the space of which is isomorphic to the space of linear operators.

A Matrix as a Tensor Let $E = \mathbb{R}^n$, $F = \mathbb{R}^p$, $x \in E$, $y \in F$, and $A \in \mathcal{L}(F, E)$:

$$F \xrightarrow{\;A\;} \quad E$$

$$y \longrightarrow x = Ay.$$

One can associate with A a bilinear form on $F \times E$ (hence a tensor) denoted A too by

$$F \times E \xrightarrow{\;A\;} \quad \mathbb{R}$$

$$(y, x) \longrightarrow A(y, x),$$

with

$$A(y, x) = \langle Ay, x \rangle. \tag{B.4.1}$$

Tensor Product of Two Vectors Let $a \in E$ and $b \in F$. The tensor product of a and b, denoted $a \otimes b$, is defined as the matrix

$$a \otimes b = a \, b^{\mathrm{T}} \quad \in \mathcal{L}(F, E). \tag{B.4.2}$$

If $a = (a_1, \ldots, a_n)$ and $b = (b_1, \ldots, b_p)$, it can be visualized as

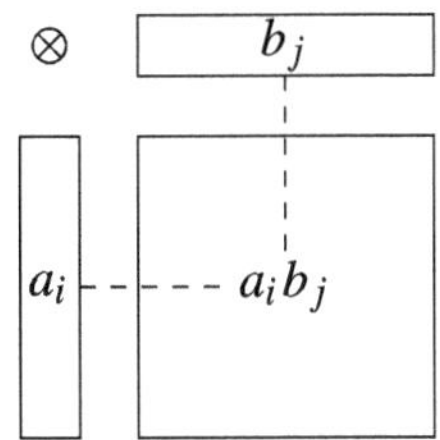

We have

$$(a \otimes b)y = \langle b, y \rangle a. \tag{B.4.3}$$

Indeed

$$\begin{aligned}
(a \otimes b)y &= (a\, b^{\mathsf{T}})\, y \\
&= a\, (b^{\mathsf{T}}\, y) \\
&= \langle b, y \rangle\, a.
\end{aligned}$$

Its associated bilinear form is

$$(a \otimes b)(x, y) = \langle a, x \rangle \langle b, y \rangle. \tag{B.4.4}$$

Indeed, $(a \otimes b)(x, y) = \langle (x, (a \otimes b)y \rangle = \langle x, \langle b, y \rangle a \rangle = \langle a, x \rangle \langle b, y \rangle$.
$a \otimes b$ is said to be of rank one or to be an elementary matrix. Indeed, $\mathrm{im}(a \otimes b) = \mathbb{R}\, a$.
It can be shown that the vector space spanned by elementary matrices is the space
of matrices in $\mathcal{L}(F, E)$. It is denoted $E \otimes F$, and

$$\mathcal{L}(F, E) \simeq E \otimes F.$$

For example,

$$\mathbb{R}^n \otimes \mathbb{R}^p \simeq \mathbb{R}^{n \times p}.$$

Decomposition of a Matrix on a Basis Let $(e_i)_i$ with $1 \le i \le n$ be basis of $\mathbb{R}^n$
and $(f_j)_j$ with $1 \le j \le p$ a basis of $\mathbb{R}^p$. Then

$$\left(e_i \otimes f_j \right)_{i,j}, \qquad \text{with} \quad \begin{cases} 1 \le i \le n \\ 1 \le j \le p \end{cases}$$

is a basis for $\mathbb{R}^n \otimes \mathbb{R}^p$. Indeed, $e_i \otimes f_j$ in this basis has zero everywhere but at the
intersection of row i and column j, and if $A = (a_{ij})_{i,j}$,

$$A = \sum_{i=1}^{n} \sum_{j=1}^{p} a_{ij}\, e_i \otimes f_j.$$

If $(f_j)_j$ is an orthonormal basis,

$$A = \sum_{j=1}^{p} A f_j \otimes f_j, \tag{B.4.5}$$

because

$$\begin{aligned}
Af_j &= \left(\sum_{i=1}^{n}\sum_{k=1}^{p} a_{ik}\, e_i \otimes f_k\right) f_j \\
&= \sum_{i=1}^{n}\sum_{k=1}^{p} a_{ik}\, (e_i \otimes f_k)\, f_j \\
&= \sum_{i=1}^{n}\sum_{k=1}^{p} a_{ik}\, \langle f_k, f_j\rangle e_i \\
&= \sum_{i=1}^{n}\sum_{k=1}^{p} a_{ik}\, \delta_k^j e_i \\
&= \sum_{i=1}^{n} a_{ij}\, e_i.
\end{aligned}$$

If $(f_j)_j$ is not orthonormal, it is possible to show with the same calculation that

$$A = \sum_{j=1}^{p} Af_j^* \otimes f_j, \tag{B.4.6}$$

where $(f_j^*)_j$ is the dual basis of (f_j), i.e., for any j, k, $\langle f_k, f_k^* \rangle = \delta_j^k$.

Rank of a Matrix We recall that the rank of a matrix is the dimension of its image space:

$$\operatorname{rank} A = \dim(\operatorname{im} A).$$

Then, the rank r of a matrix is the smallest integer $r \in \mathbb{N}$ such that there exists a decomposition of A as a sum of r matrices of rank one:

$$A = \sum_{i=1}^{r} y_i \otimes v_i, \qquad \text{with} \quad \begin{cases} A & \in \mathbb{R}^{n \times p} \\ y_i & \in \mathbb{R}^n \\ v_i & \in \mathbb{R}^p. \end{cases}$$

The family $(v_i)_i$ for $1 \leq i \leq r$ is a basis of $\operatorname{im} A$. If this basis is orthonormal, one can write

$$\|A\|^2 = \sum_{i=1}^{r} \|y_i\|^2. \tag{B.4.7}$$

Indeed,

$$\begin{aligned}
\|A\|^2 &= \langle A, A\rangle \\
&= \left\langle \sum_{i=1}^{r} y_i \otimes v_i \,,\, \sum_{j=1}^{r} y_j \otimes v_j \right\rangle \\
&= \sum_{i,j=1}^{r} \langle y_i \otimes v_i\rangle \langle y_j \otimes v_j\rangle \\
&= \sum_{i,j=1}^{r} \langle y_i, y_j\rangle \langle v_i, v_j\rangle \\
&= \sum_{i=1}^{r} \langle y_i, y_i\rangle \qquad \text{because} \quad \langle v_i, v_j\rangle = \delta_i^j \\
&= \sum_{i=1}^{r} \|y_i\|^2.
\end{aligned}$$

SVD of a Matrix Let $A \in \mathbb{R}^n \otimes \mathbb{R}^p$. Its SVD (Singular Value Decomposition) is written (if it is full rank, and $n \geq p$)

$$A = U\, \Sigma\, V^{\mathrm{T}},$$

with

$\rightarrow$ $U \in \mathbb{R}^{n \times n}$ orthonormal
$\rightarrow$ $V \in \mathbb{R}^{p \times p}$ orthonormal
$\rightarrow$ $\Sigma \in \mathbb{R}^{n \times p}$ diagonal

Such decomposition exists and is unique (see, e.g., [SS90, p.30]). The diagonal elements of Σ are called the singular values. The columns of U (respectively, V) are called the left (respectively, right) singular vectors. The matrix Σ is decomposed blockwise as

$$\Sigma = \begin{pmatrix}
\sigma_1 & 0 & \dots & 0 \\
0 & \ddots & \ddots & \vdots \\
\vdots & \ddots & \ddots & 0 \\
0 & \dots & 0 & \sigma_p \\
\vdots & \dots\dots & & 0 \\
\vdots & & & \vdots \\
0 & \dots\dots & & 0
\end{pmatrix},$$

with

$$\sigma_1 \geq \sigma_2 \geq \ldots \geq \sigma_p \geq 0.$$

A is invertible iff $\sigma_p > 0$. The rank of A is the number of nonzero elements σ_i.
SVD can be written

$$A = \sum_{i=1}^{p} \sigma_i \, u_i \otimes v_i \tag{B.4.8}$$

with $(\sigma_i)_i$ being the singular values of A, u_i the i-th column of U, and v_i the i-th
column of V. Many classical calculations can be written with SVD. For example,

$$\begin{cases} A^{\mathrm{T}} & = \displaystyle\sum_{i=1}^{p} \sigma_i \, v_i \otimes u_i \\ A^{-1} & = \displaystyle\sum_{i=1}^{p} \frac{1}{\sigma_i} \, u_i \otimes v_i. \end{cases}$$

Then, because U and V are orthonormal,

$$\|A\|^2 = \sum_{i=1}^{p} \sigma_i^2. \tag{B.4.9}$$

Indeed,

$$\begin{aligned} \|A\|^2 &= \langle A, A \rangle \\ &= \langle U \Sigma V^{\mathrm{T}}, U \Sigma V^{\mathrm{T}} \rangle \\ &= \langle \Sigma V^{\mathrm{T}}, \Sigma V^{\mathrm{T}} \rangle \\ &= \langle \Sigma, \Sigma \rangle \\ &= \|\Sigma\|^2 \\ &= \sum_{i=1}^{p} \sigma_i^2. \end{aligned}$$

The best rank r approximation of A, with $r < p$, is A_r given by

$$A_r = \sum_{i=1}^{r} \sigma_i \, u_i \otimes v_i \tag{B.4.10}$$

and

$$\|A - A_r\|^2 = \sum_{i=r+1}^{p} \sigma_i^2. \tag{B.4.11}$$

This is the basis of PCA where the $(v_i)_i$ are the principal axes and the $(y_i)_i$ with $y_i = \sigma_i u_i$ are the principal components.

Some Simple Calculations Let E, F, G be three vector spaces on the same field $\mathbb{R}$, $a \otimes b \in E \otimes F \simeq \mathcal{L}(F, E)$, and $b' \otimes c \in F \otimes G \simeq \mathcal{L}(G, F)$, sketched by

$$G \xrightarrow{b' \otimes c} F \xrightarrow{a \otimes b} E.$$

Then, $(a \otimes b)(b' \otimes c) \in E \otimes G \simeq \mathcal{L}(G, E)$ with

$$(a \otimes b)(b' \otimes c) = \langle b, b' \rangle \, a \otimes c. \tag{B.4.12}$$

Indeed

$$\begin{aligned}
(a \otimes b)(b' \otimes c) &= (a\, b^{\mathrm{T}})(b'\, c^{\mathrm{T}}) \\
&= a\, (b^{\mathrm{T}} b')\, c^{\mathrm{T}} \\
&= \langle b, b' \rangle\, a\, c^{\mathrm{T}} \\
&= \langle b, b' \rangle\, a \otimes c.
\end{aligned}$$

Let $M \in E \otimes F$, visualized as

$$G \xrightarrow{b' \otimes c} F \xrightarrow{M} E.$$

Then

$$M\,(b' \otimes c) = Mb' \otimes c. \tag{B.4.13}$$

Indeed, $M\,(b' \otimes c) \in E \otimes G \simeq \mathcal{L}(G, E)$ and

$$\begin{aligned}
M(b' \otimes c) &= M(b'\, c^{\mathrm{T}}) \\
&= (Mb')\, c^{\mathrm{T}} \\
&= Mb' \otimes c.
\end{aligned}$$

Similarly, let $Q \in F \otimes G$, visualized as

$$G \xrightarrow{Q} F \xrightarrow{a \otimes b} E.$$

Then

$$(a \otimes b)\, Q = a \otimes Q^{\mathrm{T}} b. \tag{B.4.14}$$

Indeed,

$$
\begin{aligned}
(a \otimes b)\, Q &= (a\, b^{\mathrm{T}})\, Q \\
&= a\, (b^{\mathrm{T}}\, Q) \\
&= a\, (Q^{\mathrm{T}} b)^{\mathrm{T}} \\
&= a \otimes Q^{\mathrm{T}} b.
\end{aligned}
$$

B.5 Random Projection

Vector spaces of very large dimension exhibit properties we are not familiar with from our geometric intuition developed in $\mathbb{R}^2$ or $\mathbb{R}^3$. For our purpose of linear dimension reduction, this leads to the approaches relying on so-called random projection, which is gently introduced here. This is only the tip of an iceberg, addressing issues in measure concentration and the geometry of Banach spaces. We restrict ourselves here on what is focused toward linear dimension reduction, i.e., SVD with random projection.

Isometries, Orthogonal Matrices, and Rotations

Let us have a point cloud in $\mathbb{R}^p$. Random projection consists in choosing randomly a subspace E_k of small dimension, say k, in $\mathbb{R}^p$, and project the point cloud on E_k. One question is to characterize and choose randomly such a subspace. It can be done by choosing randomly an orthonormal basis of E_k. Randomness is realized by choosing vectors with a uniform measure on the sphere (the Haar measure) with constraints of orthogonality. This in turn can be done by choosing randomly a rotation, i.e., a $p \times p$ matrix in special orthogonal group, and keeping the first k columns only.

Let $n \in \mathbb{N}$. The set of invertible matrices in $\mathbb{R}^{n \times n}$ is the linear group of $\mathbb{R}^n$, denoted $\mathbb{GL}(n)$.

$\diamond$ A matrix $A \in \mathbb{R}^{n \times n}$ is orthogonal if

$$A A^{\mathrm{T}} = A^{\mathrm{T}} A = \mathbb{I}_n. \tag{B.5.1}$$

It is invertible, and

$$A^{-1} = A^{\mathrm{T}}. \tag{B.5.2}$$

The set of all orthogonal matrices on $\mathbb{R}^n$ is called the orthogonal group and is denoted $\mathbb{O}(n)$:

$$\mathbb{O}(n) = \{A \in \mathbb{R}^{n \times n} \mid AA^{\mathsf{T}} = A^{\mathsf{T}}A = \mathbb{I}_n\}. \tag{B.5.3}$$

It is one of the classical compact Lie groups in $\mathbb{GL}(n)$. We have

$$\det A = \pm 1. \tag{B.5.4}$$

The Special Orthogonal Group $\mathbb{SO}(n)$ is the set of orthogonal matrices with determinant equal to one

$$\mathbb{SO}(n) = \{A \in \mathbb{O}(n) \mid \det A = 1\}. \tag{B.5.5}$$

Its elements are called rotations. The set of orthogonal matrices such that $\det A = -1$ is often denoted $\mathbb{SO}^-(n)$. Its elements are called reflections.

◇ A matrix A is orthogonal if, and only if,

→ Its columns form an orthonormal basis of $\mathbb{R}^n$.
→ It acts as an isometry on $\mathbb{R}^n$, i.e.,

$$\forall x, y \in \mathbb{R}^n, \quad \langle Ax, Ay \rangle = \langle x, y \rangle. \tag{B.5.6}$$

Concentration of the Measure on the Sphere

Concentration of the measure is an unexpected result in spaces of very large dimension about the global variations of a function whose local variations are kept small. The control of local variations is given by Lipschitz property. It is an immense domain, which is merely touched here for measure concentration on the sphere which will be useful for a proof of Johnson-Lindenstrauss lemma.

Lipschitz Function Let E, F be two metric spaces (here, we will consider distances being associated with a norm in a finite dimensional vector space). A function

$$E \xrightarrow{\ f\ } F$$

is said Lipschitz if there exists a constant C such that

$$\forall x, y \in E, \quad d(f(x) - f(y)) \leq Cd(x, y). \tag{B.5.7}$$

A linear map is a Lipschitz function. Indeed, let $L \in \mathcal{L}(E, F)$ be a linear map. The spectral norm $\|.\|_{\mathrm{sp}}$ is defined as

$$\|L\|_{\mathrm{sp}} = \max_{x \neq 0} \frac{\|Lx\|}{\|x\|}, \tag{B.5.8}$$

where $\|.\|$ is the Frobenius norm. Hence, for all x, $\|Lx\| \leq \|L\|_{\mathrm{sp}} \|x\|$. Letting $x \to x - y$ leads to

$$\|Lx - Ly\| = \|L(x - y)\| \leq \|L\|_{\mathrm{sp}} \|x - y\|. \tag{B.5.9}$$

Measure Concentration on the Sphere Let $\mathbb{S}^{n-1}$ denote the sphere in $\mathbb{R}^n$:

$$\mathbb{S}^{n-1} = \{x \in \mathbb{R}^n \mid \|x\| = 1\}. \tag{B.5.10}$$

Let

$$\mathbb{S}^{n-1} \xrightarrow{\ f\ } \mathbb{R}$$

be a Lipschitz function with constant C. Let m be the median of this function on the sphere (i.e., a value such that for a random vector x on the sphere, the probability that $f(x) \geq m$ is larger than or equal to $1/2$, as well as the probability that $f(x) \leq 1/2$). Let x be a uniformly chosen random vector on the sphere. Then, Levy's lemma, or measure concentration on the sphere, asserts that, for any $t > 0$,

$$\mathbb{P}(|f(x) - m| \geq tC) \leq 2 \exp{-(n - 2)t^2}. \tag{B.5.11}$$

If $n \to \infty$, the quantity on the r.h.s. $\to 0$ whatever t. Hence, for very large dimensions $n > N \gg 1$, the function f is essentially equal to its median because

$$\forall t > 0, \quad \forall \epsilon > 0, \quad \exists N \in \mathbb{N} \quad \text{s.t.} \quad \forall n > N, \quad \mathbb{P}(|f(x) - m| \geq tC) \leq \epsilon. \tag{B.5.12}$$

$\diamond$ Here is an example. Let us select randomly a vector $a \in \mathbb{S}^{n-1}$. It will be called the "north pole" of the sphere. Let us define the map

$$\mathbb{S}^{n-1} \xrightarrow{\ f\ } \mathbb{R}$$
$$x \longmapsto \langle a, x \rangle. \tag{B.5.13}$$

It is a linear form and hence is Lipschitz. We have

$$|f(x) - f(y)| = |\langle a, x \rangle - \langle a, y \rangle| = |\langle a, x - y \rangle| \leq \|a\| \|x - y\| = \|x - y\|. \tag{B.5.14}$$

Hence the constant is $C = 1$. The median is $m = 0$. Indeed, let us consider the hyperplane H orthogonal to a. All points on the sphere in the upper half-space defined by it (the part that contains a) are such that $f(x) \geq 0$, and those in the lower half-space which contains $-a$ are such that $f(x) \leq 0$ ($f(x) = 0$ for the points on the equator, defined as $\mathbb{S}^{n-1} \cap H$). We then have

$$\mathbb{P}(|f(x)| \geq t) \leq 2 \exp -(n-2)t^2. \tag{B.5.15}$$

Then, $f(x) = \langle a, x \rangle$ is zero almost everywhere (recall that $m = 0$), and nearly all points are close to the equator. This is an asymptotic result: The fraction of points out of the t-neighborhood of the equator becomes negligible for very large dimensions only. For this fraction to be lower than a given ϵ for a given t, one must have

$$n > 2 + \frac{1}{t^2} \operatorname{Log} \frac{2}{\epsilon} \approx \frac{1}{t^2} \operatorname{Log} \frac{2}{\epsilon}. \tag{B.5.16}$$

Selecting $t = \epsilon = 10^{-2}$ yields $n \approx 5.3 \times 10^4$.

Notes and References A very clear and accessible introduction to the concentration of measure is [Led01]. A basic example already known to Borel and mentioned in the introduction of [Led01] is the geometric interpretation of the law of large numbers, given as follows. Let us consider the hypercube $\mathbb{K}_n = [0, 1]^n \subset \mathbb{R}^n$ and H be its intersection with the hyperplane orthogonal to the diagonal from $(0, \ldots, 0)$ to $(1, \ldots, 1)$ at $(1/2, \ldots, 1/2)$. Let H_t be the set of points in $\mathbb{K}_n$ at distance $\leq t$ from H. Then, if μ is the uniform measure in $\mathbb{K}_n$ (noticing that $\mu(\mathbb{K}_n) = 1$), i.e., $d\mu = dx_1 \ldots dx_n$, then $\mu(H_{t\sqrt{n}}) \to 1$ when $n \to \infty$. All points in $\mathbb{K}_n$ are concentrated in the t-neighborhood of H for n sufficiently large. Their projection on the diagonal is concentrated on the segment $[1/2 - t\sqrt{n}, 1/2 + t\sqrt{n}]$. This is the law of large numbers.

The Johnson-Lindenstrauss Lemma

The Johnson-Lindenstrauss lemma is about a good surprise for linear dimension reduction: Loosely speaking, for a random point cloud X of size m in a high dimension space $\mathbb{R}^n$, and an accuracy ϵ, there exist a dimension k and a subspace $E \subset \mathbb{R}^n$ of dimension k such that the distances between the projections of points in X on E approximate the distances between the points in X with accuracy ϵ. Moreover, if the space E is chosen at random, the probability that this accuracy is ϵ is nonzero. The lower bound for k depends on m, not on n. This is understandable, because a point cloud of m points with Euclidean distances can be embedded into a m-dimensional space. So, the lemma is often given with $m = n$. For k to be significantly smaller than m when ϵ is small, m must be very, very large. In brief, a point cloud of m points is concentrated with accuracy ϵ on a vector subspace of dimension $\approx 8 \operatorname{Log} m/\epsilon^2$.

◇ Let us first give some notations:

$\quad\rightarrow\quad$ $n \in \mathbb{N}$ and $\epsilon \in [0, 1/2]$.
$\quad\rightarrow\quad$ $X = (x_i)_i$ is a point cloud of m points with $1 \le i \le m$, $x_i \in \mathbb{R}^n$.
$\quad\rightarrow\quad$ Let

$$k \ge \frac{4 \operatorname{Log} m}{\epsilon^2/2 - \epsilon^3/3}.$$

Then, there exists a linear map

$$\mathbb{R}^n \xrightarrow{\;f\;} \mathbb{R}^k$$

such that

$$\forall i, j, \quad (1-\epsilon)\|x_i - x_j\| \le \|f(x_i) - f(x_j)\| \le (1+\epsilon)\|x_i - x_j\|. \tag{B.5.17}$$

This is an amazing lemma, which tells that in very large dimensions ($n \gg 1$), a cloud X of n points in $\mathbb{R}^n$ is sharply concentrated on a vector subspace of dimension $k \approx \operatorname{Log} n$.

The demonstration sketched here is borrowed from [DG03]. We assume $m = n$.

◇ Let $x \in \mathbb{R}^n$ be a point chosen uniformly on the surface of the sphere. Therefore, we choose $y \in \mathbb{R}^n$ with $y_i \sim \mathcal{N}(0, 1)$, and let $x = y/\|y\|$. Let E_k be the vector subspace of $\mathbb{R}^n$ spanned by the first k vectors of the standard basis of $\mathbb{R}^n$. It can be any vector subspace of dimension k: It suffices to choose the basis of $\mathbb{R}^n$ accordingly. Let z be the projection of x on E_k, and $\zeta = \|z\|^2$. Let $\mu = \mathbb{E}(\zeta)$.

◇ The first step in the demonstration, detailed in [DG03], is to show that ζ is sharply concentrated around $\mu = k/d$. Indeed, we have

$$\mathbb{P}(\zeta \le (1-\epsilon)\mu) \le \exp -\frac{k(\epsilon^2/2 - \epsilon^3/3)}{2}. \tag{B.5.18}$$

The demonstration of this lemma relies on the observation that the map $x \mapsto \zeta$ is Lipshitz, with $L = 1$, and on an application of measure concentration on the sphere. If $k \ge \frac{4 \operatorname{Log} n}{\epsilon^2/2 - \epsilon^3/3}$, this yields

$$\mathbb{P}(\zeta \le (1-\epsilon)\mu) \le \frac{1}{n^2}. \tag{B.5.19}$$

It can be shown similarly that

$$\mathbb{P}(\zeta \ge (1+\epsilon)\mu) \le \frac{1}{n^2}. \tag{B.5.20}$$

So

$$\mathbb{P}\left(\zeta \notin [(1-\epsilon)\mu,\ (1+\epsilon)\mu]\right) \leq \frac{2}{n^2}. \tag{B.5.21}$$

$\diamond$ Let us define the map

$$x \xrightarrow{\ f\ } \sqrt{\tfrac{n}{k}}\, z.$$

Let us consider a pair (x_i, x_j) and $x = x_i - x_j$. Then

$$\mathbb{P}\left(1 - \epsilon \leq \frac{\|f(x_i) - f(x_j)\|^2}{\|x_i - x_j\|^2} \leq 1 + \epsilon\right) \geq \frac{2}{n^2}. \tag{B.5.22}$$

$\diamond$ The next step is to use the so-called *union bond*, known also as Boole's inequality. If we have n events $(A_i)_{1 \leq i \leq n}$, then

$$\mathbb{P}\left(\bigcup_i A_i\right) \leq \sum_i \mathbb{P}(A_i).$$

From this, we can deduce, from $\overline{A \cup B} = \overline{A} \cap \overline{B}$,

$$\mathbb{P}\left(\bigcap_i \overline{A_i}\right) \geq 1 - \sum_i \mathbb{P}(A_i).$$

Let us consider the events A_{ij} defined by

$$A_{ij} := \left\{ \frac{\|f(x_i) - f(x_j)\|^2}{\|x_i - x_j\|^2} \notin [1 + \epsilon, 1 + \epsilon] \right\}.$$

We have

$$\bigcap_{1 \leq i < j \leq n} \overline{A_{ij}} = \left\{ \forall\, i, j, \quad 1 - \epsilon \leq \frac{\|f(x_i) - f(x_j)\|^2}{\|x_i - x_j\|^2} \leq 1 + \epsilon \right\},$$

which is the event that none of the pair (x_i, x_j) has a distortion greater than ϵ which is the property to be shown. There are $n(n-1)/2$ events A_{ij}, with

$$\forall\, (i, j), \quad \mathbb{P}(A_{ij}) \leq \frac{2}{n^2}.$$

Then

$$\sum_{1 \le i < j \le n} \mathbb{P}(A_{ij}) \le \frac{n(n-1)}{2} \frac{2}{n^2} = 1 - \frac{1}{n}, \tag{B.5.23}$$

and

$$\mathbb{P}\left(\bigcap_{i<j} \overline{A_{ij}}\right) \ge 1 - \sum_{i<j} \mathbb{P}(A_{ij}) \ge \frac{1}{n} > 0.$$

$\diamond$ Let us recall that E_k has been chosen at random. Let us denote by π the probability that, for any subspace E_k chosen at random, the distortion of the point cloud projected on E_k is less than ϵ. We have

$$\pi \ge \frac{1}{n}.$$

Let us consider q subspaces E_k, all chosen at random. The probability $p(q)$ that at least for one of them, the distortion is less than ϵ is

$$p(q) = 1 - (1 - \pi)^q.$$

As $\pi \ge 1/n$,

$$p(q) \ge 1 - \left(1 - \frac{1}{n}\right)^q.$$

So, as

$$\lim_{q \to \infty} p(q) = 1,$$

there exists one subspace E_k for dimension k with a distortion less than ϵ.

Notes and References The Johnson-Lindenstrauss lemma is the starting point of many developments in the domain of dimension reduction and random algorithms in linear algebra. It is often referred to as the *flattening lemma*, because it flattens a point cloud. The lemma given here has been borrowed from [Mat08] and [DG03]. The demonstration is sketched in [Mat08] and detailed in [Mat02, section 15.2] or [DG03], which are similar and rely on the same observations, from which it has been borrowed and to which the reader may refer to all details omitted here. [Mat02] gives also some historical notes and points to several surveys on JL lemma and its utilization in a diversity of algorithms. Let us mention however that, for having $k \ll n$ with the given bound $k \ge \frac{4 \operatorname{Log} n}{\epsilon^2/2 - \epsilon^3/3}$ for small ϵ, n must be huge. For example, if $\epsilon = 10^{-2}$ and $n = 10^6$, then $k > n$. If $n = 10^7$, then $k \ge 1.28 \times 10^6$. For $\epsilon = 10^{-1}$, this drops to 1.2×10^4 for $n = 10^7$, and $k \ge 8.6 \times 10^3$ for $n = 10^5$. The threshold value of k is very sensitive to ϵ, as $k = O(\epsilon^{-2} \log n)$. However, luckily, the projection is often of very good quality for much lower dimensions $k' < k$. There is a link between MDS and the choice of E_k fulfilling the conditions of JL lemma. One can say loosely that classical MDS is finding the space with the smallest average distortion, whereas JL is about finding a space with a bounded distortion whatever the pair of points, which is the ℓ^∞ norm for the distortion on pairs. Finally, it is tempting to propose an algorithm for finding a bounded distortion space knowing that the probability of the condition being fulfilled for a randomly chosen space is greater than $1/n$: All we have to do is try randomly until the condition is fulfilled. However, we have seen that the probability of a success after q trials is in $1 - (1 - 1/n)^q$. If $q = n$, this leads to $q(n) \sim 1 - e^{-1} \approx 0.63$. So, the chance of success after n trials (and n is very large) is about two thirds. It is not an efficient way.

Bibliography

[ACD+22] E. Agullo, O. Coulaud, A. Denis, M. Faverge, A. Franc, J.-M. Frigerio, N. Furmento, A. Guilbaud, E. Jeannot, R. Peressoni, F. Pruvost, and S. Thibault, Task-based randomized singular value decomposition and multidimensional scaling. Research Report RR-9482, Inria Bordeaux - Sud Ouest; Inrae - BioGeCo, 2022

[And58] T.W. Anderson, *An introduction to Multivariate Statistical Analysis* (John Wiley & Sons, New York, 1958)

[Bas94] A. Basilevsky, *Statistical Factor Analysis and Related Methods: Theory and Applications* (John Wiley & Sons, New York, 1994)

[BCF+18] P. Blanchard, P. Chaumeil, J.-M. Frigerio, F. Rimet, F. Salin, S. Thérond, O. Coulaud, A. Franc, A geometric view of Biodiversity: scaling to metagenomics. Research Report RR-9144, INRIA; INRA, 2018

[BEK90] R. Bhatia, L. Elsner, G. Krause, Bounds for the variation of the roots of a polynomial and the eigenvalues of a matrix. Linear Algebra Appl. **142**, 195–209 (1990)

[Bel57] R. Bellman, *Dynamic Programming* (Princeton University Press, Princeton, 1957)

[Ben73a] J.-P. Benzecri, *L'Analyse des Données; tome 2: l'analyse des correspondances* (Dunod, 1973)

[Ben73b] J.-P. Benzecri, *L'Analyse des Données, tome 1: la taxinomie* (Dunod, 1973)

[Ber87] M. Berger, *Geometry, I.* Universitext (Springer, Berlin, 1987)

[BG05] I. Borg, P.J.F. Groenen, *Modern Multidimensional Scaling*, 2nd edn. Springer Series in Statistics (Springer, Berlin, 2005)

[Bha82] R Bhatia, Analysis of spectral variation and some inequalities. Trans. Am. Math. Soc. **272**(1), 323–331 (1982)

[Bha87] R. Bhatia, *Perturbation Bounds for Matrix Eigenvalues*, volume 162 of Research Notes in Mathematics Series (Pitman, 1987)

[Bha07] R. Bhatia, Spectral variation, normal matrices, and Finsler geometry (2007). http://www.isid.ac.in/~statmath/eprints

[Bis06] C.M. Bishop, *Pattern Recognition and Machine Learning* (Springer, Berlin, 2006)

[Bla17] P. Blanchard, *Fast hierarchical algorithms for the low-rank approximation of matrices with applications to materials physics, geostatistics and data analysis*. Ph.D. Thesis, University of Bordeaux, 2017

[BLM13] S. Boucheron, G. Lugosi, P. Massart, *Concentration Inequalities—A Nonasymptotic Theory of Independence* (Oxford University Press, Oxford, 2013)

[BM01] E. Bingham, H. Mannila, Random projection in dimensionality reduction: applications to image and text data, in *Proceedings of the Seventh ACM SIGKDD International Conference on Knowledge Discovery and Data Mining* (2001), pp. 245–250

[CC80] C. Chatfield, A.J. Collins, *Introduction to Multivariate Analysis* (Chapman & Hall, 1980)

[CC01] T.F. Cox, M.A.A. Cox, *Multidimensional Scaling*, 2nd edn., volume 88 of Monographs on Statistics and Applied Probability (Chapman & Hall, 2001)

[Cla87] A. Claret, *Contribution au problème de l'approximation factorielle d'un tableau de données*. Ph.D. Thesis, Université des Sciences et techniques du Languedoc, 1987.

[CP79] F. Cailliez, J.-P. Pagès, *Introduction à l'Analyse des Données* (S.M.A.S.H. Editions, 1979)

[CST00] N. Cristianini, J. Shawe-Taylor, *An Introduction to Support Vector Machines and Other Kernel-Based Learning Methods* (Cambridge University Press, Cambridge, 2000)

[DD07] S. Dray, A.B. Dufour, The ade4 package: implementing the duality diagram for ecologists. J. Stat. Softw. **22**(4), 1–20 (2007)

[DG03] S. Dasgupta, A. Gupta, An elementary proof of a theorem of Johnson and Lindenstrauss. Random Struct. Algorithms **22**(1), 60–65 (2003)

[DlCH11] O. De la Cruz, S. Holmes, The duality diagram in data analysis: examples of modern applications. Ann. Appl. Stat. **5**(4), 2266–2277 (2011)

[dSP10] C. de Seguin Pazzis, *Invitation aux formes quadratiques* (Calvage & Mounet, Paris, 2010)

[Els82] L. Elsner, On the variation of the spectra of matrices. Linear Algebra Appl. **47**, 127–138 (1982)

[EP90] B. Escofier, J. Pagès, *Analyse Factorielles simple et multiples* (Dunod, Paris, 1990)

[EY36] K. Eckart, G. Young, The approximation of a matrix by another of lower rank. Psychometrika **1**(3), 211–218 (1936)

[Fra92] A. Franc, *Etude Algébrique des multitableaux: apports de l'algèbre tensorielle*. Ph.D. Thesis. Université Montpellier 2, 1992

[Gar11] D.J.H. Garfiled, *Clifford Algebras: An Introduction*, volume 78 of London Mathematical Society Student Texts (Cambridge University Press, Cambridge, 2011)

[GB06] M. Greenacre, J. Blasius (eds.), *Multiple Correspondence Analysis and Related Methods* (Chapman & Hall, 2006)

[GH08] A. Galántai, C.J. Hegedüs, Perturbation bounds for polynomials. Numer. Math. **109**, 77–100 (2008)

[Git85] R. Gittins, *Canonical Analysis: A Review with Applications in Ecology*, volume 12 of Biomathematics (Springer, Berlin, 1985)

[GM19] F.P. Greenleaf, S. Marques, *Linear Algebra I* (American Mathematical Society, 2019)

[GM20] F.G. Greenleaf, S. Marques, *Linear Algebra, II* (American Mathematical Society, 2020)

[Gol84] L. Goldfarb, A unified approach to pattern recognition. Pattern Recognit. **17**(5), 575–582 (1984)

[Gow85] J.C. Gower, Properties of Euclidean and non-Euclidean distance matrices. Linear Algebra Appl. **67**, 81–97 (1985)

[Gre84] M. Greenacre, *Theory and Applications of Correspondence Analysis* (Academic Press, New York, 1984).

[Hen62] P. Henrici, Bounds for iterates, inverses, spectral variation and fields of values of nonnormal matrices. Nume **4**, 24–40 (1962)

[Hil74] M.O. Hill, Correspondence analysis: a neglected multivariate method. J. Roy. Stat. Soc. Ser. C (Applied Statistics) **23**(3), 340–354 (1974)

[HJ12] R.A. Horn, C.R. Johnson, *Matrix Analysis*, 2nd edn. (Cambridge, 2012).

[HLP17] F. Husson, S. Lê, J. Pagès, *Exploratory Multivariate Analysis by Examples Using R*, 2nd edn. (CRC Press, Boca Raton, 2017)

[HMT11] N. Halko, P.G. Martinsson, J.A. Tropp, Finding structure with randomness: probabilistic algorithms for constructing approximate matrix decompositions. SIAM Rev. **53**(2), 217–288 (2011)

[Hol92] J.A. Holbrook, Spectral variation of normal matrices. Linear Algebra Appl. **174**, 131–144 (1992)

[HTF09] T. Hastie, R. Tishibani, J. Friedman, *The Elements of Statistical Learning*, 2nd edn. Springer Series in Statistics (Springer, Berlin, 2009)

[Ive92] B. Iversen, *Hyperbolic Geometry*, volume 25 of London Mathematical Society Student Texts (Cambridge University Press, Cambridge, 1992)

[Ize08] A.J. Izenman, *Modern Multivariate Statistical Techniques* (Springer, New York, 2008)

[Jac91] J.E. Jackson, *A User's Guide to Principal Components* (Wiley, London, 1991)

[JC16] I.T. Jolliffe, J. Cadima, Principal component analysis: a review and recent developments. Philos. Trans. R. Soc. A **374**(2065), 20150202 (2016)

[JL84] W.B. Johnson, G. Lindenstrauss, Extension of Lipschitz mapping into a Hilbert space. Contemp. Math. **26**, 189–206 (1984)

[Jol02] I.T. Jolliffe, *Principal Component Analysis*, 2nd edn. (Springer, Berlin, 2002)

[Lau98] M. Laurent, A connection between positive semidefinite and Euclidean distance matrix completion problems. Linear Algebra Appl. **273**, 9–22 (1998)

[Led01] M. Ledoux, *The Concentration of Measure Phenomenon*, volume 89 of Mathematical Surveys and Monographs (American Mathematical Society, 2001)

[LMF82] L. Lebart, A. Morineau, J.-P. Fénelon, *Traitement des données statistiques* (Dunod, Paris, 1982)

[LMP00] L. Lebart, A. Morineau, M. Piron, *Statistique exploratoire multidimensionnelle* (Dunod, Paris, 2000)

[LMT77] L. Lebart, A. Morineau, N. Tabard, *Techniques de la description statistique* (Bordas - Dunod, 1977)

[LSBB91] J.D. Lebreton, R. Sabatier, G. Banco, A.M. Bacou, *Principal Component and Correspondence Analyses with Respect to Instrumental Variables: An Overview of Their Role in Studies of Structure - Activity and Species - Environment Relationships*, chapter 2 (Springer, Berlin, 1991), pp. 85–114

[LV07] J.A. Lee, M. Verleysen, *Nonlinear Dimensionality Reduction* (Springer, New York, 2007)

[Mat02] J. Matoušek, *Lectures on Discrete Geometry*, volume 212 of Graduate Texts in Mathematics (Springer, Berlin, 2002)

[Mat08] J. Matoušek, On variants of the Johnson-Lindenstrauss lemma. Random Struct. Algorithms **33**. 142–156 (2008)

[Mec19] E. Meckes, *The Random Matrix Theory of the Classical Compact Groups*. Cambridge Tracts in Mathematics, vol. 218 (Cambridge University Press, Cambridge, 2019)

[Mec20] E. Meckes, The eigenvalues of random matrices. IMAGE (2020), pp. 9–22. https://arxiv.org/pdf/2101.02928.pdf

[MKB79] K.V. Mardia, J.T. Kent, J.M. Bibby, *Multivariate Analysis*. Probability and Mathematical Statistics (Academic Press, New York, 1979)

[Mur12] K.P. Murphy, *Machine Learning: A Probabilistic Perspective* (MIT Press, Cambridge, 2012)

[MZ24] M. Meila, H. Zhang, Manifold learning: what, how, and why. Annu. Rev. Stat. Appl. **11**(1), 393–417 (2024)

[NG07] O. Nenadić, M. Greenacre, Correspondence analysis in R, with two- and three-dimensional graphics: the CA package. J. Stat. Softw. **20**(3), 1–12 (2007)

[Par18] E. Paradis, Multidimensional scaling with very large datasets. J. Comput. Graph. Stat. **27**(4), 935–939 (2018)

[PCY79] J.-P. Pagès, F. Cailliez, Y. Escoufier, Analyse factorielle: un peu d'histoire et de géométrie. Revue de Statistiques Appliquées **27**(1), 5–28 (1979)

[PD05] E. Pekalska, R.P.W. Duin, *The Dissimilarity Representation for Pattern Recognition. Foundations and Applications* (World Scientific, Singapore, 2005)

[Pea01] K. Pearson, On lines and planes of closest fit to systems of points in space. Lond. Edinb. Dublin Philos. Mag. J. Sci. **2**(11), 559–572 (1901)

[Rao64] C.R. Rao, The use and interpretation of principal component analysis in applied research. Sankhya **26**(4), 329–368 (1964)

[Rao73] C.R. Rao, *Linear Statistical Infernence and its Applications*. Wiley Series in Probability and Mathematical Statistics, 2nd edn. (Wiley, London, 1973)

[Sab84] R. Sabatier, Quelques généralisations de l'Analyse en Composantes Principales de Variables Instrumentales. Stat. Ann. Donn. **9**(3), 75–103 (1984)

[Sap90] G. Saporta, *Probabilités, Analyse de Données et Statistique* (Editions Technip, 1990)

[Sch38] I.J. Schoenberg, Metric spaces and positive definite functions. Trans. Am. Math. Soc. **44**(3), 522–536 (1938)

[Sch60] N. Schatten, *Norms Ideals of Completely Continuous Operators* (Springer, Berlin, 1960)

[Sch97] J.R. Schott, *Matrix Analysis for Statistics* (John Wiley & sons, New York, 1997)

[SS90] G.W. Stewart, J. Sun, *Matrix Perturbation Theory* (Academic Press, New York, 1990)

[SSBD14] S. Shalev-Shwartz, S. Ben-David, *Understanding Machine Learning—From Theory to Algorithms* (Cambridge University Press, Cambridge, 2014)

[Str19] G. Strang, *Linear Algebra and Learning from Data* (Wellesley Cambridge Press, 2019)

[TB99] M.E. Tipping, C.M. Bishop, Probabilistic principal component analysis. J. R. Statist. Soc. B **61**(3), 611–622 (1999)

[TDD+18] J. Thioulouse, S. Dray, A.-B. Dufour, A. Siberchicot, T. Jombart, S. Pavoine, *Multivariate Analysis of Ecological Data with ADE4* (Springer, Berlin, 2018)

[Tor52] W.S. Torgerson, Multidimensional scaling: I. Theory and method. Psychometrika **17**(4), 401–419 (1952)

[TY85] M. Tenenhaus, F.W. Young, An analysis and synthesis of Multiple Correspondence Analysis, Optimal Scaling, Dual Scaling, Homogeneity Analysis and other methods for quantifying multivariate categorical data. Psychometrika **50**(1), 91–119 (1985)

[Vem04] S.S. Vempala, *The Random Projection Method*, volume 65 of DIMACS Series in Discrete Mathematics and Theoretical Computer Sciences (American Mathematical Society, 2004)

[VMS16] R. Vidal, Y. Ma, S. S. Sastry, *Generalized Principal Component Analysis* (Springer, Berlin, 2016)

[Wan12] J. Wang, *Geometric Structure of High-Dimensional Data and Dimensionality Reduction* (Springer & Higher Education Press, 2012)

[Wol87] S. Wold, Principal component analysis. Chemometrics Intell. Lab. Syst. **2**, 37–52 (1987)

MIX
Papier aus verantwortungsvollen Quellen
Paper from responsible sources
FSC® C105338

If you have any concerns about our products,
you can contact us on
ProductSafety@springernature.com

In case Publisher is established outside the EU,
the EU authorized representative is:
Springer Nature Customer Service Center GmbH
Europaplatz 3, 69115 Heidelberg, Germany

Printed by Libri Plureos GmbH
in Hamburg, Germany